Efficiencies from Applying a Rotational Equipping Strategy

Christopher G. Pernin, Edward Wu, Aaron Martin,
Gregory Midgette, Brendan See

Prepared for the United States Army
Approved for public release; distribution unlimited

RAND ARROYO CENTER

The research described in this report was sponsored by the United States Army under Contract No. W74V8H-06-C-0001.

Library of Congress Control Number: 2011925854

ISBN: 978-0-8330-5200-1

Published 2011 by the RAND Corporation
1776 Main Street, P.O. Box 2138, Santa Monica, CA 90407-2138
1200 South Hayes Street, Arlington, VA 22202-5050
4570 Fifth Avenue, Suite 600, Pittsburgh, PA 15213-2665
RAND URL: http://www.rand.org/
To order RAND documents or to obtain additional information, contact
Distribution Services: Telephone: (310) 451-7002;
Fax: (310) 451-6915; Email: order@rand.org

Preface

The Army has adopted and implemented a rotational strategy over the past few years, and many facets of its operations have adapted to support that strategy. One area still coming in line with the strategy is how the Army equips and modernizes its units. RAND Arroyo Center was asked to consider how an equipping strategy tailored to a rotational force might affect the equipment that the Army needs to buy and the near- and far-term effects that it might have on its budgets.

The purpose of this document is to show how a rotational equipping strategy (RES) for some pieces of equipment and some units might reduce total equipment needed and therefore provide considerable opportunities for cost savings in the near- and far-term defense spending plans. This document and the findings therein should be of interest to those planning and budgeting for the force and those interested in force development issues.

This research was sponsored by LTG Robert Lennox, Deputy Chief of Staff, G-8. It was conducted in RAND Arroyo Center's Force Development and Technology Program. RAND Arroyo Center, part of the RAND Corporation, is a federally funded research and development center sponsored by the United States Army.

The Project Unique Identification Code (PUIC) for the project that produced this document is RAND10486.

For more information on RAND Arroyo Center, contact the Director of Operations (telephone 310-393-0411, extension 6419; fax 310-451-6952; email Marcy_Agmon@rand.org), or visit Arroyo's website at http://www.rand.org/ard.html.

Contents

Preface ... iii
Figures ... vii
Tables ... ix
Summary ... xi
Acknowledgments ... xix
Abbreviations ... xxi

CHAPTER ONE
The Army's Rotational Equipping Strategy ... 1
Introduction ... 1
Can a Rotational Equipping Strategy Save Costs? ... 4
About This Report ... 5

CHAPTER TWO
Demands on the Rotational Force ... 7
Demand Is an Important Input to Equipping a Rotational Army ... 7
Possible Demand Assumptions ... 10
Force Structure–Based Demand Assumptions ... 12
Similarities to Current Force Structure ... 15
Summary ... 16
About the Next Chapter ... 17

CHAPTER THREE
Applying the Rotational Equipping Strategy ... 19
Applying the RES ... 19
Applicable Army Units for Rotational Equipping ... 20

Appropriate Equipment Types for Rotational Equipping 20
Application of Rotational Equipping to Select Equipment and Units..... 26
The Rotational-High Level 27
The Rotational-Low Level 27
Application to Various Items 28
Budgetary Effects of Implementing the RES 29
Conclusions About the Budget 32

CHAPTER FOUR
Conclusions 35
Policy Decisions 36
Implications 36

APPENDIX
Modeling the Rotational Equipping Strategy 39

Bibliography 59

Figures

S.1. Deployed and CONUS Inventory During a Snapshot in 2007 ... xii
1.1. Deployed and CONUS Inventory During a Snapshot in January 2007 ... 3
1.2. The Rotational Equipping and Modernization Model ... 5
2.1. Theater-Based Brigade Combat Team Demand ... 11
2.2. Force Structure–Based Demand ... 14
3.1. Candidate Equipment for the RES ... 23
3.2. Apache Inventories over Time Under Different Equipping Alternatives ... 31
A.1. REMM Demand Model Overview ... 46
A.2. Simulated Demand for M109 over 15 Years ... 48
A.3. REMM Supply Model Overview ... 50
A.4. Major End Item Numbers Needed as a Function of Demand CABs ... 53
A.5. Major End Item Numbers Needed as a Function of Demand HBCTs ... 54
A.6. Major End Item Numbers Needed as a Function of Number of AC CABs ... 55
A.7. Major End Item Numbers Needed as a Function of Number of AC HBCTs ... 55
A.8. Average Days Deployed Early per Deployment, per Combat Brigade, as a Function of the Demand for CABs ... 56
A.9. Average Days Deployed Early per Deployment, per Combat Brigade, as a Function of the Demand for HBCTs ... 56

Tables

S.1. Rotational-High and Rotational-Low Numbers for Some Major End Items xiv
1.1. Secretary of Defense Guidance on Cost Savings (billions of dollars per year) 4
2.1. Steady-State Supply Based on Aggregate Force Structure 15
3.1. Rotational Units 21
3.2. Major Item Procurement Program Expenditures for FYs 2011–2015 25
3.3. Procurement and Expenditures for Apaches, Chinooks, and Paladins 26
3.4. Rotational-High and Rotational-Low Numbers for Some Major End Items 29

Summary

Background and Problem

The past near decade of conflict in Iraq and Afghanistan has changed the way the force is managed. To meet the demands of protracted conflict in those theaters, the Army has adopted a rotational deployment strategy based on the Army Force Generation (ARFORGEN) model. In this model, units rotate through various levels of readiness, with a portion (approximately one-third) immediately available for deployments as part of a full-spectrum force. With this new strategy, and unlike the tiered readiness strategy of years past, all units pass through high-readiness phases during a portion of their ARFORGEN cycle.

While many of the Army's policies have adapted to the ARFORGEN model, the equipping policies still largely reflect Cold War tradition to provide active, Reserve, and National Guard units with 100 percent of their authorized equipment at all times during the ARFORGEN cycle. Since units are rotating through various states of readiness—and at times can be multiple years from any deployment—the utility of such an equipment policy is questionable.

Data from a snapshot of equipment locations illustrate the potential problem (Figure S.1). In 2007, somewhere between only 10 and 40 percent of available high-end equipment was deployed during a period of time in which the United States was engaged in one of the largest sustained deployments in recent times. From an efficiency standpoint, the Army may be able to find considerable efficiencies by bringing its equipment fill levels more in line with other unit readiness levels.

Figure S.1
Deployed and CONUS Inventory During a Snapshot in 2007

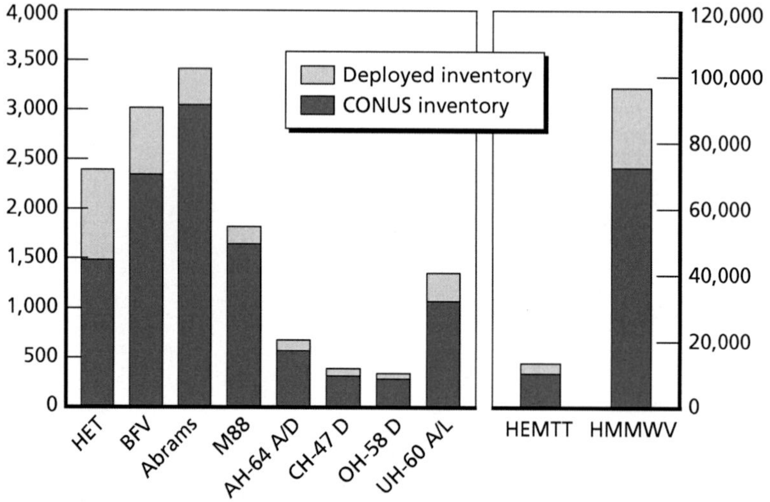

SOURCE: W. M. Solis, *Defense Logistics: Preliminary Observations on the Army's Implementation of Its Equipment Reset Strategy*, Washington, D.C.: United States Government Accountability Office, 2007.
NOTES: BFV = Bradley Fighting Vehicle, HEMTT = heavy expanded mobility tactical truck, HET = heavy equipment transporter, HMMWV = high-mobility multipurpose wheeled vehicle.
RAND *MG1092-S.1*

Analyzing the Rotational Equipping Strategy

In order to analyze the potential impacts of reduced equipment and what that might mean to the force, we constructed a simulation of unit deployments, based on ARFORGEN, to determine equipment inventories to support steady-state and surge operations. The model evaluates current or planned force structure and equipment,[1] demands for deployments, and policies governing rotating units and equipment to determine the amount of equipment necessary to meet Army

[1] For this analysis, we winnowed the list of items and units for which an RES would work best to those with sufficient number (>300), those with appreciable value (>$100 million in total replacement cost), and those items low in stock compared with authorizations. This provides a starting point for the items to which an RES would best be applied and should be updated as more information becomes available as the strategy is executed.

goals. From the modeling, we are able to determine how far equipment levels might be suppressed while still meeting training and deployment needs. To do so, we defined two alternative strategies for equipping units through the ARFORGEN cycle. These we termed the *rotational-low* and the *rotational-high* levels of aggregate equipment.[2]

The outputs of our analysis show how many major end items the Army would require for each strategy. As shown in Table S.1, the quantities of many of the Army's most prominent systems might feasibly be reduced up to 25 percent through an RES while still meeting training and potential surge demands for forces. For some of the equipment, the reductions are much smaller, owing to the force structure, broad utility of the equipment, and ratio of specific types of active component (AC) and reserve component (RC) units.

There are many implications of adopting an RES. First, the reduced equipping levels mean that during steady-state operations, the Army would have to set specific levels for equipment fills for major equipment that are in line with the training and potential surge needs of the force. These numbers will change from item to item, and thus a nuanced look at each item is necessary. The examples we show are based on discussion with Army planners and can be updated as more detailed information is available.

Second, during surges of forces, the Army would need to ready units and the support infrastructure to transition considerable equipment from units in the United States to units deployed in theater, as well as equipment from units returning to home to other units in the continental United States. The RES is predicated on equipment being resident somewhere in the Army for surge deployments, and, therefore, a sharing of equipment (with all attending cultural and logistical implications) will need to be worked out.

---

[2] The rotational-high level plans for a reduced amount of equipment in early phases of ARFORGEN and preserves those fill rates regardless of surge requirements (e.g., equipment cannot be pulled from units that are not deployed). The rotational-low level plans for a further-reduced amount of equipment to some minimum amount necessary for training purposes and allows units that are deploying to "borrow" equipment from units not deployed during surge situations.

Table S.1
Rotational-High and Rotational-Low Numbers for Some Major End Items

System	Total (MTOE + TDA + APS)	Rotational-Low Number (% of Total)	Rotational-High Number (% of Total)
AH-64D	772	550 (71%)	688 (89%)
CH-47	404	368 (91%)	398 (98%)
UH-60	1,467	1,351 (92%)	1,446 (98%)
HH-60	570	522 (92%)	561 (98%)
M109	616	454 (73%)	567 (92%)
M1A1	1,100	878 (80%)	1,028 (93%)
M1A2	832	664 (80%)	774 (93%)
M2A2	998	753 (75%)	905 (91%)
M2A3	953	778 (82%)	889 (93%)
M2A2ODS	351	246 (70%)	316 (90%)
M7 BFIST	315	227 (72%)	276 (88%)
M3A2	564	450 (80%)	520 (92%)
M3A3	402	314 (78%)	368 (92%)

NOTES: The data above only include combat aviation brigades, theater aviation brigades, fires brigades, and heavy brigade combat teams for all equipment. APS = Army prepositioned stock, MTOE = modified table of organization and equipment, TDA = tables of distribution and allowances.

Third, the analysis in this report entailed a detailed modeling effort and thus rests on a number of assumptions that will need to be understood, monitored, and updated as circumstances warrant. The analysis is highly dependent on a few of those assumptions, notably the demand assumptions (as detailed in Chapter Two) and rules of how ARFORGEN is executed (see the appendix for additional information on how various input parameters drive the outputs of the model). Other assumptions about rates of equipment lost in combat and transit times are also important and thus amenable to additional analysis and strategies to reduce or control risks. Nonetheless, careful consideration of the assumptions underpinning the analysis can offer insights into how an RES would affect equipping and modernizing the Army.

Fourth, an RES is a significant change in how the Army equips units—and, therefore, a potentially significant change in how soldiers view units and the Army. While this analysis does not detail the cultural changes necessary to implement such a rotational strategy, it will be necessary to include such considerations as the Army moves forward.

What Might the Rotational Equipping Strategy Mean to the Budget?

The assumptions detailed in the previous section, therefore, can be considered policy decisions that the Army faces, should it fully execute an RES. There are three assumptions described in this report that bear repeating:

1. The Army will have to decide what demand they will assume for planning purposes—or whether their force structure and rules of ARFORGEN will drive that number.
2. The Army will have to decide how much equipment each unit will get for each piece of equipment during the early phases of ARFORGEN.
3. The rules of ARFORGEN—particularly those associated with deployment and dwell times—need to be formally adopted for equipment planning purposes.

In this report, we assumed answers for all of these decisions through rationales built from senior decisionmakers within the Army, such as the Army's Chief of Staff, and found considerable cost savings potentially available.

Rotational equipping provides opportunities to reduce the total amount of some equipment in the force. The equipment most amenable to an RES is typically the most expensive, has sufficient numbers for swapping among units, can easily be moved and integrated from unit to unit, and is not needed in large quantities for training. Many of the Army's major end items are available for rotational equipping.

The effect that smaller inventories have on the budget is threefold. First, with a smaller target number for a given fleet, the Army can **divest of those systems most in need of repair and recapitalization** until the rotational level is met. The possible cost avoidance from this kind of strategy needs to be determined but would have to be weighed against any increased maintenance costs that could be incurred because of the higher usage rates inherent in a smaller fleet.[3]

Second, there are potential medium- and longer-term savings from **reducing the total number of systems needing upgrades**. The Army's modernization programs are steeped in upgrades to major systems, such as the aircraft and tracked vehicles. If a rotational strategy that allows the Army to lower the overall number of systems needed is adopted, those upgrade programs could reduce the tail ends of the programs to save money. (Note that reducing near-term upgrades is also a possibility but would sacrifice some near-term capability.) As an example, the CH-47, UH-60M, and AH-64 upgrade programs total over $30 billion in fiscal year (FY) 2012 and beyond. Finding a means of reducing the total stock necessary to upgrade will have profound impacts in the mid- to long-term budget, and our analysis shows that $5 to $10 billion of savings is likely. Based on the analysis shown in this report, we estimate that reducing inventories for the Paladin, CH-47, and AH-64 to rotational-low levels of equipment may avoid between $1.7 and $4.0 billion in upgrade costs over the lifespan of their current modernization programs. This depends on whether new builds are forgone or not, in addition to reduction of current inventories. Alternatively, reducing inventories to rotational-high levels may save between zero and $1.1 billion over the same period.[4]

Lastly, a smaller inventory of major end items can provide options for **reducing or even eliminating new purchases in the near term**. Since current Army plans call for the procurement of replacement air-

3 Analysis has not yet been performed to determine which case costs more for specific platforms: smaller fleets used intensely or larger fleets used sporadically.

4 All numbers are in then-year dollars and are approximately $0.3 billion for ten D to F conversions in the Chinook, $0.5 billion in the Paladin upgrades if 89 of the upgrades are avoided, and $0.9 billion in FY 2023 and FY 2024 if 85 Apache Block III upgrades are avoided.

craft, vehicles, and weapons, a smaller overall Army requirement may offer the opportunity to eliminate or reduce those purchases. As an example, the two new combat aviation brigades that the Army is standing up based on the 2010 *Quadrennial Defense Review* (QDR) will lead to new purchases of both CH-47s and AH-64s, for a combined cost of approximately $4.5 billion. Our analysis of a few examples shows how reducing the overall Army authorization levels can reduce near-term procurements totaling billions of dollars across the Five Years Defense Plan. Specifically, we estimate an avoidance of up to $4 billion for the CH-47 and AH-64 fleets if rotational-low levels are adopted.[5]

---

5 This breaks down as approximately $2.1 billion for the 56 new-build Apaches and another $2.1 billion for the new Chinook purchases.

Acknowledgments

The project team would like to thank our sponsor, LTG Robert Lennox, for his support of this effort and his feedback during interim project briefings. We also would like to thank MG Thomas Spoehr for his input on the project and for giving us access to his directorate to gather information and discuss the finer points of an RES from a force development standpoint. In particular, we were fortunate to have discussions with COL Randy Rotte, who provided feedback and data on the aviation examples discussed in this report.

Our project monitor within the G-8 QDR office was Tim Muchmore, who provided significant input early in the project and provided access to information and people and critical feedback throughout the project. Within that office, we thank MAJ Momoevi Tawake for his support in data collection and coordination. We would also like to thank the various equipment experts within the Army who were generous with their time while explaining how equipment is considered within the Army. This includes CW4 Graham Ward for information on the aviation portfolio, Carl Wiley for information on Apaches, CW5 Bobby McQuaid for details on Apache life-cycle operations, and CPT Jerry McCulley and MAJ Carlos Rivera for information on tanks and Bradleys.

Within RAND, we were fortunate to have detailed discussions with colleagues well versed in logistical issues surrounding equipment levels, including Matt Lewis, Rick Eden, and their teams working on other equipping issues.

Abbreviations

AB2	Apache Block II
AB3	Apache Block III
AC	active component
AH	attack helicopter
APS	Army prepositioned stock
ARB	armed reconnaissance battalion
ARFORGEN	Army Force Generation
ARNG	Army National Guard
BCT	brigade combat team
BFSB	battlefield surveillance brigade
BOG	boots on the ground
CAB	combat aviation brigade
CDU	critical dual use
CONUS	continental United States
DoD	Department of Defense
FA brigade	field artillery brigade
FMTV	family of medium tactical vehicles

FY	fiscal year
FYDP	Five Years Defense Plan
GMLRS	guided multiple launch rocket system
HBCT	heavy brigade combat team
HQ	headquarters
IBCT	infantry brigade combat team
ISR	intelligence, surveillance, and reconnaissance
JTRS	joint tactical radio system
MCO	major combat operation
MEB	maneuver enhancement brigade
MEEL	mission-essential equipment list
MTOE	modified table of organization and equipment
MTW	major theater war
OEF	Operation Enduring Freedom
OIF	Operation Iraqi Freedom
QDR	Quadrennial Defense Review
RC	reserve components
REMM	Rotational Equipping and Modernization Model
RES	rotational equipping strategy
SBCT	Stryker brigade combat team
SECDEF	Secretary of Defense
SME	subject matter expert
TAB	theater aviation brigade
TDA	tables of distribution and allowances

TO&E	table of organization and equipment
TPE	theater-provided equipment

CHAPTER ONE

The Army's Rotational Equipping Strategy

Introduction

The past near decade of conflict in Iraq and Afghanistan has changed the way the force is managed. To meet the demands of protracted conflict in those theaters, the Army has adopted a rotational deployment strategy based on what is referred to as the Army Force Generation (ARFORGEN) model. ARFORGEN calls for rotating forces through three phases—Ready, Available, and Reset—in order to provide a steady-state supply of forces to combatant commanders.[1] Thus, the Army's current position, as articulated by the Chief of Staff of the Army, is to provide a portion (approximately one-third) of the force as available for immediate deployments.

The rotational strategy is similar in some ways to the tiered readiness strategies used prior to 2000, except that instead of units being static in their readiness, with some ready and others in various degraded levels of readiness, units are rotated through readiness phases over time. In the new rotational strategy, all units will have the opportunity to be available for deployments for some portion of time during their rotational cycle.

Many parts of the Army have adapted to the ARFORGEN model. In late 2009, the Army released a new equipping strategy that explained how it might change its equipping practices to align with

[1] More information on the ARFORGEN model can be found in the 2010 Army Posture Statement (2010 Army Posture Statement, "Addendum F, Army Force Generation [ARFORGEN], the Army's Core Process," Army.mil, last modified March 2, 2010).

and exploit the rotational force generation strategy.[2] The new equipping strategy challenges many past notions of equipping and, when implemented, would address some of the key areas affecting the rotational force.[3] Broadly speaking, the new equipping strategy aims to provide soldiers with the equipment they need to accomplish their mission as they progress through the cyclic readiness model (ARFORGEN).[4] The Army's strategy is not necessarily a reduction or appreciable change in the amount of equipment it would need to purchase, but rather is an aligning of its equipping strategy with its new readiness strategy.

The Army's new equipping strategy evolved in response to current protracted conflicts in Iraq and Afghanistan. However, current practices continue to reflect a Cold War tradition because the Army continues to strive to provide active, Reserve, and National Guard units with 100 percent of the equipment listed in its tables of organization and equipment (TO&E). Since units only need a full authorization of equipment while deployed or in the Available phase of ARFORGEN, and they can train and prepare in early phases of ARFORGEN at lower levels of equipment, the Army might not have to equip all forces to the full TO&E authorization at all times.

Data from a snapshot of equipment location (Figure 1.1) illustrate the perceptual problem that equipping a rotational force to the TO&E at all times creates. In 2007, somewhere between only 10 and 40 percent of available high-end equipment was deployed. After taking out minimum amounts for training, reset, and other friction, this leaves a substantial capital investment resident in the United States that is not being used during one of the largest sustained deployments in recent times. Some of the continental United States–based (CONUS-based) equipment is unnecessarily with units that are in early phases of ARFORGEN, and that equipment may be reduced, depending on what it is being used for.

---

2 Headquarters, Department of the Army, *The Army Equipping Strategy*, white paper, Washington, D.C., 2009.

3 As of September 2010, the rotational strategy released in November 2009 had not been completely implemented, though some efforts to execute the strategy are under way.

4 Headquarters, Department of the Army, 2009, p. 2.

Figure 1.1
Deployed and CONUS Inventory During a Snapshot in January 2007

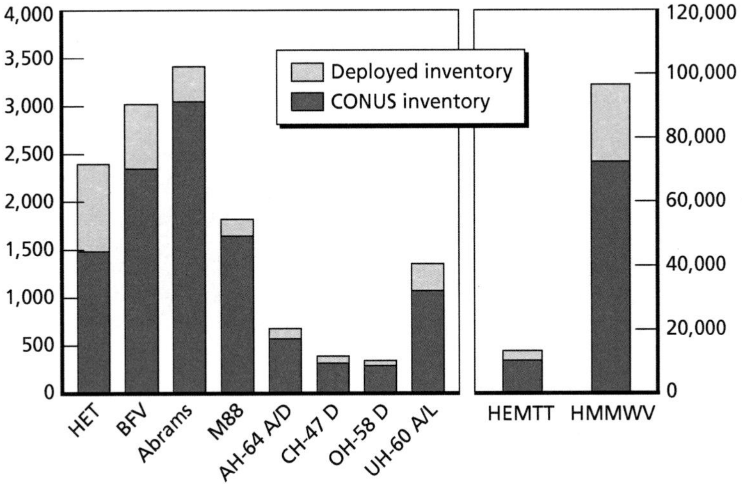

SOURCE: W. M. Solis, *Defense Logistics: Preliminary Observations on the Army's Implementation of Its Equipment Reset Strategy*, Washington, D.C.: United States Government Accountability Office, 2007.
NOTES: BFV = Bradley Fighting Vehicle, HEMTT = heavy expanded mobility tactical truck, HET = heavy equipment transporter, HMMWV = high-mobility multipurpose wheeled vehicle.
RAND *MG1092-1.1*

The new equipping strategy allows for changes in how the Army equips the reserve components (RC). During recent contingencies in Iraq and Afghanistan, the RC have been used as an operational reserve and is fully integrated into the ARFORGEN process. Units and individuals of the RC are routinely activated and deployed to meet theater demands. In fact, by a RAND estimate based on Department of Defense (DoD) data, the RC have provided nearly 32 percent of deployments to theater. It thus cannot simply be equipped with older, out-of-date equipment from the active component (AC), as was often done in the past.[5]

The new equipping strategy also focuses on dynamically equipping units to the specific missions in their upcoming rotational cycle.

---

[5] Timothy Bonds, Dave Baiocchi, and Laurie McDonald, *Army Deployments to OIF and OEF,* Santa Monica, Calif.: RAND Corporation, DB-587-A, 2010.

The current conflicts have been steeped in the development and procurement of new and advanced weaponry; vehicles; and command, control, communications, computers, intelligence, surveillance, and reconnaissance systems to fight in a complex environment against adaptive threats. The challenge has been to identify and equip units specifically to the fights to which they are deploying, and not to the fights they may have been designed or originally equipping to fight. Whereas past equipping was based on a unit's modified table of organization and equipment (MTOE), the equipping strategy for the current fight is often based on tailored mission-essential equipment lists (MEELs) and theater-provided equipment (TPE), where some units are falling in on equipment they had never seen before.

Can a Rotational Equipping Strategy Save Costs?

The Army's operational tempo has necessitated considerable spending, in both base and supplemental budgets, to meet wartime requirements for new equipment. In fact, the overall Army procurement budget has increased nearly fourfold in the past eight years to procure new, advanced materiel to prosecute two protracted conflicts;[6] reset the force from the increased use of equipment; and transform to a modular, brigade-centric force. More recently, with the drawdown in Iraq and budget pressures mounting, the Secretary of Defense (SECDEF) has directed the services to find efficiency savings that can be applied to more pressing requirements. The Army's portion of the requested savings is about $28 billion from FY 2012 through FY 2016 (see Table 1.1).

Table 1.1
Secretary of Defense Guidance on Cost Savings (billions of dollars per year)

	FY 2012	FY 2013	FY 2014	FY 2015	FY 2016
Reductions per service	2	3	5.3	8	10

6 In 2001 procurement was $12.5 billion (FY 2010 dollars) and in 2009 reached its peak at $46.5 billion, which is an increase of $34 billion, or 273 percent.

While the SECDEF has indicated support for continuing proportionately large defense budgets in coming years, the downturn in the economy and historical precedence for postwar spending make considering budget declines in the near and mid-term a necessity. The question from an equipment standpoint is: Can a new rotational equipping strategy (RES) provide cost savings?

About This Report

This report illustrates how the Army might execute an RES, how far the concept might be pushed, and what effects that strategy might have in the near- and far-term Army budgets. This report deals directly with implementing the RES for an ARFORGEN-based Army, with the objective to define options for equipping a rotational force at least cost while preserving capacity and capability.

For this study, we built a fast-running analytic model to simulate the deployment of forces and allocation of equipment to units over long-term steady-state and surge deployments. The model, depicted graphically in Figure 1.2, takes force structure, demands, and rules of how forces rotate through the ARFORGEN cycle and produces minimum and maximum equipping levels. This model relies on a number of important policy decisions that have not been made yet—and pro-

Figure 1.2
The Rotational Equipping and Modernization Model

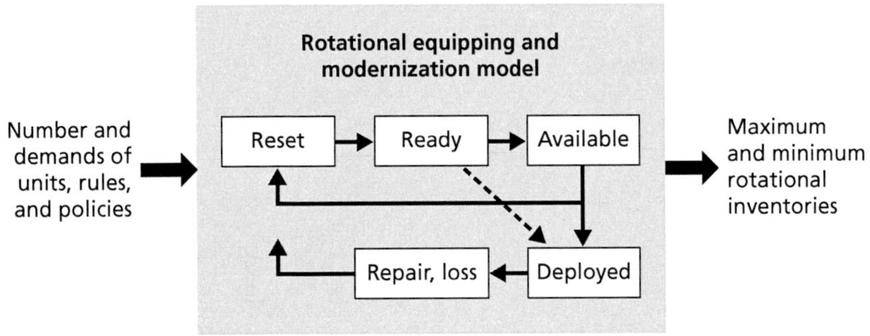

RAND *MG1092-1.2*

vides a basis for making those decisions in the coming years. We apply the model to a number of pieces of equipment to show how an RES would affect major items and service budgets.

Chapter Two describes the role of demand in defining required equipping. Our analysis assumes that current plans for force structure remain relevant, although the method we use could also be used with excursions to explore how changes to force structure or planning requirements and demands might impact equipping needs in the future.

In Chapter Three, we apply the model and method to a number of major end items and further show how changes to the inventory from adopting an RES might be applied to the Army's near- and far-term budgets to cut costs. The potential for finding efficiencies in the budget from such a change in equipping strategy is then related to the SECDEF budget guidance to determine the extent to which the strategy might be useful in the coming years.

Chapter Four provides some conclusions based on the analysis, and an appendix describes the model we developed. The model uses the Army's current (or near-term) force structure, planning factors for the demand of forces, and rules and policies for a rotational Army to determine overall equipping requirements. A future report will detail the model.

CHAPTER TWO

Demands on the Rotational Force

Accurately predicting the future demand for Army deployments is impossible, but creating a framework for considering the potential range of these deployments is vital to ensuring that the Army possesses an appropriate amount of equipment. In this chapter, we describe the significance of developing a model for deployment demands, explain alternative ways to model demand, describe our approach, and highlight potential problems with deployment-based demand assumptions.

Demand Is an Important Input to Equipping a Rotational Army

During the decades-long Cold War, threat-based, or scenario-based, planning provided force developers a planning construct from which to work. This method for sizing a force offered a unique threat and clear mission goals with a definable (although uncertain) sizing requirement and has been accompanied by analytical processes and techniques to adjudicate the sufficiency of current force structures and capabilities. Over many decades, major conflicts, defined as major theater wars (MTWs), have dominated force planning. Across the military, the individual services place different importance on one conflict or collections of conflicts over others, and, over time, each service has vacillated between treating such scenarios as real threats or notional planning constructs. The Army, by and large, has relied on MTW scenarios to plan and modernize its force and, in doing so, assumed demands to mobilize and potentially employ the entire Army. Using this type of demand construct to examine required equipment levels, we find that the Army would need enough equipment to fully stock all units for full deployment. An alternative approach to equipping is to capitalize on the Army's current rotational readiness strategy and equip to a unit's position in the ARFORGEN cycle to determine whether certain efficiencies exist. With an RES, the Army assumes that some proportion of units will not deploy during steady-state operations. For those units not deployed, the unit's requirement for equipment is driven not by how much equipment they require for wartime operations, but rather by training needs and amounts necessary for use by other units during a surge. Thus, defining the demand for such forces—both steady state and for surges—becomes an important assumption in devising a sustainable equipping strategy. A demand-driven approach to the RES works by applying an appropriate amount of equipment to units throughout the ARFORGEN cycle and then examining what happens to that equipment as units rotate to and from deployments. This method suggests that a deployment model, even if imperfect, is essential because it provides an ability to consider operational issues that affect the stock of equipment the Army needs.

Since the basic concept of the RES is that the portion of equipment required by a unit increases as that unit progresses through ARFORGEN, it may appear that a model for demand is unnecessary for examining the stock of equipment required by the Army, as long as each unit has all of the equipment it requires when it enters the phase of the cycle in which it may deploy. However, this analytic method fails to recognize several operational factors that influence the utilization of equipment and the pace of a rotational cycle. The ARFORGEN readiness model and other rotational readiness cycles must provide flexibility for dealing with changing demand levels; ARFORGEN deals with this explicitly by stating that units in the Train-Ready phase must be deployable in 90 to 180 days, and units in the Ready phase can deploy 180 days after notification.[1] Failing to account for these surge conditions in the readiness cycle could lead to underestimating the stock of

---

[1] George W. Casey, "The Army of the 21st Century," *Army Magazine*, October 2009, pp. 25–40.

equipment the Army needs, should a situation occur in which it needs to deploy forces from the Train-Ready or Reset pools. A deployment demand assumption should include the consideration of surge conditions by providing an estimate of the additional stock of equipment needed to meet potential future threats.

Modeling deployment demands also provides the opportunity to consider more subtle equipping issues that arise during real deployments. These challenges include:

- **Transport time to and from theater.** Equipment typically takes several weeks to move to and from a unit's home station and to an overseas theater of operations. This time also varies for different modes of transportation. Transportation costs and potential impacts on requirements become much more important in protracted environments in which units take regular deployments abroad.
- **Equipment attrition.** Equipment has a much higher probability of being destroyed or consumed during combat operations than during training and therefore needs to be considered in terms of risk to the force. Careful consideration of potential theater equipment attrition rates is required to balance the gains of reduced overall equipment numbers against the risk of equipment shortfalls in theater.
- **Increased needs for maintenance.** The Army found that equipment deployed during recent operations often requires higher levels of maintenance during and following the time deployed. Significant amounts of equipment have been pulled from units returning after deployment for reset at the units' installations or reset or recapitalization at depots.
- **Unit overlap in theater.** Army units overlap in theater for a certain period of time in order to facilitate the transition between units, which creates additional demand for equipment.
- **Dwell times.** Policies for dwell times (i.e., time at the home station) and boots on the ground (BOG) times change with different assumptions of demand and force structuring.

Modeling the deployment demands provides an opportunity to consider how these factors may affect the overall stock of equipment needed by the Army. In some cases, the equipment may be assigned to a unit but not on hand, such as when the equipment is in transport to a unit deploying to theater or when units reconstituting after deployment send a proportion of their items to depot, while in others our estimates for the overall equipment needs of the Army increase.

Including a model of deployment demands also allows for the exploration of policy levers that may reduce the total stock of equipment needed by the Army. Above we explained that there is some flexibility in the ARFORGEN cycle, as units may be required to deploy along a slower time frame than in the earlier stages in the ARFORGEN cycle. Through an exploration of the flexibility of the ARFORGEN cycle using a deployment model, we were able to examine aspects of an equipping strategy that allows for the reassignment of equipment from the units that do not deploy and those in the Reset phase of ARFORGEN. This provides an opportunity for the Army to accept equipment deficits in some units during deployment surges in order to reduce the total stock of a particular item.

Possible Demand Assumptions

There are many ways to include future deployment demands into analysis. Future demand may be constructed using data from recent deployments, a set of example deployments, or estimating the forces needed during possible future scenarios.[2] DoD typically uses a set of planning scenarios for force sizing purposes. Our demand signals are not designed to examine whether the current force size is appropriate but instead focus on estimating the stock of equipment that the Army needs, given a set of deployment demands. Recognizing that future demand is impossible to accurately predict, we used demand data

[2] In this report, we will not address future planning scenarios, which may be classified. The model we propose, however, is methodologically consistent with implementing such a demand construct.

from recent military deployments dating back to Operations DESERT SHIELD and DESERT STORM while also accounting for forces that remain committed in other countries, like Korea, as a starting point for our discussion.

Our consideration of past deployments based on military operations and commitments produced four separate theaters of operation that represent different types of deployment situations ranging from commitments in Korea to major combat operations. We used past experiences to develop a 15-year extrapolation in order to ensure multiple rotations for both the AC and RC forces through the ARFORGEN cycle.[3] Figure 2.1 highlights the demand for brigade combat teams (BCTs) for each of the individual theaters.

Theater 1 represents the current commitment to Korea, where the Army currently has one heavy brigade combat team (HBCT) that maintains a high level of readiness. Demand in theaters 2 and 3 represents two different types of steady-state operations. The demand in theater 2 represents potential demand for forces during an operation

Figure 2.1
Theater-Based Brigade Combat Team Demand

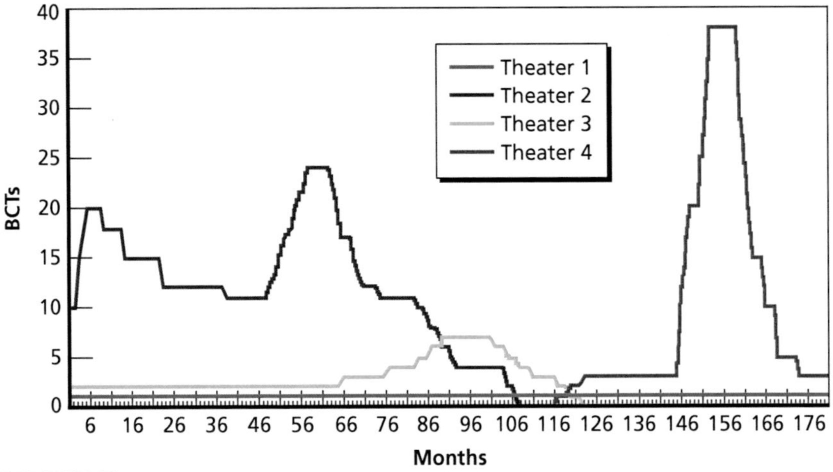

RAND *MG1092-2.1*

[3] We ran the model for 15 years to ensure that it had stabilized and included all phasing effects of unit deployments.

requiring a long-term, relatively large force level, while the demand in theater 3 represents potential demand for a smaller steady-state deployment. We allowed the demand in each of these theaters to vary in order to represent unexpected surges or lulls during steady-state operations.

Finally, theater 4 represents a major short-term commitment of Army forces for an extreme event. We examined the composition of Army forces during Operations DESERT SHIELD and DESERT STORM and found that those Army forces were composed of two corps headquarters (HQs) that were deployed along with about eight Army divisions and several separate brigades or cavalry regiments. We also examined the Army's attack aviation assets and found that the Army deployed a total of 419 attack helicopters (274 Apaches and 145 Cobras).[4] The force demanded in theater 4 is roughly the size of the force deployed during the Persian Gulf War, representing some extreme event.

There are several potential problems when modeling demand based on recent requirements. The largest potential problem is that future demands may not reflect past experiences. The future may prove to have more or less challenging situations than the past or may use previously unavailable formations.[5] One way to address this challenge is to construct demands that entail fully deploying all available forces; this would provide greater stress on the rotational cycle, ensuring that the Army has adequate equipment for a very difficult or "worst case" future.

Force Structure–Based Demand Assumptions

An alternative approach to incorporating demand based on historical experience or possible future scenarios is to input demands that fully employ the Army force structure according to the readiness expecta-

[4] U.S. Department of Defense, *Conduct of the Persian Gulf War,* Final Report to Congress Pursuant to Title V of the Persian Gulf Supplemental Authorization and Personnel Benefits Act of 1991, Washington, D.C., April 1992, pp. 283–291 and 755–760.

[5] For example, as the Army has converted to modular brigades and supporting structures, those new units may not have precedence or similar capabilities as those units deployed in previous contingencies. See Casey (2009), pp. 25–40, for additional detail.

tions of the ARFORGEN cycle. Instead of seeking to determine what future demands for Army units may actually be, this demand assumption takes the maximum expected forces that the Army could supply to all operations and then deploys them. Army leadership expects that the modular design of Army units, in both the AC and RC, cycling through the ARFORGEN process will increase the readiness of the overall force by providing a consistent supply of deployable units and reduce uncertainty about deployments for unit members.[6]

Recent Army documents and statements from Army officials identify several different goals for the progression through the ARFORGEN process. These goals describe a ratio indicating the amount of time in the Available pool against the amount of time in the Reset and Train-Ready pools; the differences among these goals are primarily determined by whether they are short- or long-term goals. The 2009 *Army Posture Statement* explains that under steady-state requirements the ARFORGEN goal for the ratio of time characterized as Ready to home station time (i.e., the ratio of BOG time to dwell time, or BOG:dwell) is 1:3 during a three-year rotation for AC forces and 1:5 for RC forces, while BOG:dwell goals change to 1:2 for AC and 1:4 for RC under surge conditions.[7] On the other hand, the Army Chief of Staff stated that the goal for FY 2011 is 1:2 for AC forces and 1:4 for RC forces.[8] The 2010 *Quadrennial Defense Review* (QDR) establishes a BOG:dwell goal of 1:2 for the AC and 1:5 for the RC.[9]

The demand based on ARFORGEN availability focuses on the unit availability expectations outlined by the Army Chief of Staff. According to his statement, the Army expects to provide one corps HQ, five division HQs, 20 BCTs, and 92,000 enablers in the Available pool at any one time. Forces in the Train-Ready and Reset pools are each roughly the same size as those in the Available pool. We con-

---

6 U.S. Department of the Army, "Addendum E: Army Force Generation Model—ARFORGEN," *2006 Posture Statement*, site last revised February 8, 2006.

7 U.S. Department of the Army, 2006.

8 Casey, 2009, p. 32.

9 U.S. Department of Defense, *Quadrennial Defense Review Report*, Washington, D.C., February 2010, p. 51.

structed a demand scenario lasting for 15 years that maximizes the possible deployments from this cycle. This demand requires a steady-state force package of 20 BCTs deployed throughout the entire period. A short-term surge of up to 40 BCTs for one year, taken from the Train-Ready pool, is available for emergency situations.[10]

To manage the challenge of determining how the headquarters and support force structures may look during this time, we applied the same type of deployment rules as we did for BCTs. We deployed the entire Available pool and included a deployment of twice that level during the surge period. Figure 2.2 highlights what this demand looks like for BCTs over the 15-year period.

Using these deployment assumptions allows us to generate estimates of the total stock of equipment the Army may need to meet a challenging deployment environment while equipping its forces accord-

Figure 2.2
Force Structure–Based Demand

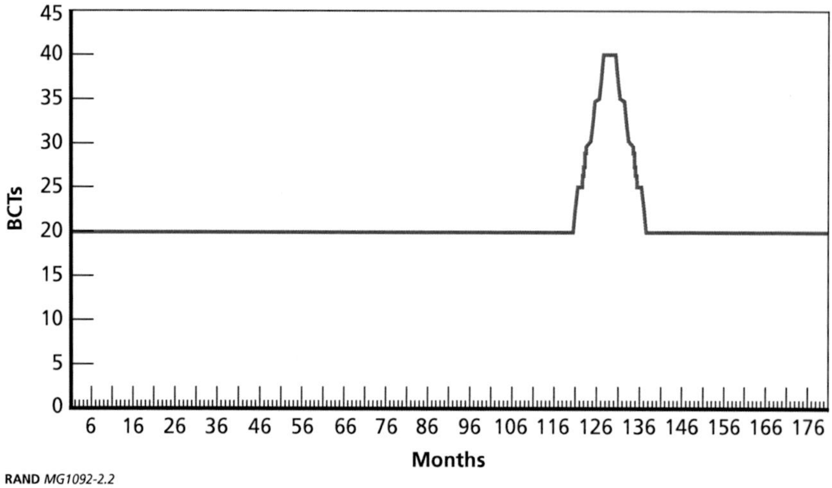

RAND *MG1092-2.2*

[10] To put this into context, a 20-BCT steady-state deployment is approximately equivalent to deployments in Iraq and Afghanistan combined in the 2007–2008 time period; the 40-BCT deployment is approximately 20 percent larger than the Army's initial deployment into Operation DESERT STORM in 1991.

ing to an RES.[11] As guidance and new information become available, updating the demand assumptions is a straightforward process. For instance, if these estimates are lower than the stock of equipment the Army currently has or expects to have, the Army may be able to reduce its stock of equipment without increasing operational risk.

Similarities to Current Force Structure

The demand assumption described above follows guidance from the Army Chief of Staff on what the Army will provide. The numbers are consistent with the force structure that is based on previously mentioned BOG:dwell ratios. As shown in Table 2.1, the planned force structure can be rotated "high" (with BOG:dwell ratios for AC of 1:2 and RC of 1:4) or "low" (with BOG:dwell ratios for AC of 1:3 and RC of 1:5) to provide approximately 20 BCTs or approximately 15 BCTs, respectively, at steady state. While these are rough estimates based on aggre-

Table 2.1
Steady-State Supply Based on Aggregate Force Structure

	Planned Number[a]		Steady-State Available (High)	Steady-State Available (Low)
Unit Type	**AC**	**RC**	**1:2 and 1:4**	**1:3 and 1:5**
Division HQ	10	8	4.9	3.8
BCT	45	28	20.6	15.9
CAB	13	8	5.9	4.6
Apache ARB[b]	18	10	8.0	6.2

NOTE: ARB = armed reconnaissance battalion, CAB = combat aviation brigade.

[a] Planned force structure comes from the QDR 2010 (U.S. Department of Defense, 2010).

[b] Planned Apache battalions assumes that medium-variant CABs are added to the force.

---

[11] It is assumed here that this supply-based deployment scenario encompasses scenarios being used for Army planning.

gate force structures, they are in line with the current guidance and provide a top-end estimate of what the force might provide, without including any friction that might exist.[12]

Similarly, the deployable amounts of the other parts of the Army's rotational force structure can be calculated. For instance, with this rough estimation, division HQs can be rotated to provide approximately five (for high) and approximately four (for low) deployed in steady-state conditions. Because of the inevitable friction that comes with so many moving parts, these are only meant to be upper bounds to the pace at which the force can rotate based on stated deployment policies.

Summary

Deployment demands are essential to estimating the equipment levels that the Army needs to implement an RES. We described several methods for estimating demands in this section and explained that we focused our modeling efforts by assuming that all available forces are deployed with a major deployment surge. This type of demand seeks to create the most challenging future the Army may expect to face, based on its current force structure and deployment goals. After considering both scenarios and force structure–based demand, we determined that the equipment levels required to fully deploy Army forces eclipse those from scenario-based demands. These levels reduce the risk that the Army would not meet its deployment requirements because of equipment shortfalls.

Some potential challenges still exist when using this demand assumption. The force structure–based demand assumption assumes that the size and composition of Army forces are appropriate for future challenges. This demand is not useful in all situations; for example, future situations may require more of one type of BCT than are avail-

---

[12] Examples of friction include overlap in units, nondeployable units (perhaps those nonrotational units permanently stationed elsewhere), and other areas, which will suppress the total amount of supply the Army might provide.

able through ARFORGEN. A model that only seeks to deploy forces based on those available cannot identify this type of situation. Additionally, the model is sensitive to the component in which forces reside because the AC and RC have different BOG:dwell goals. Migrating forces from one component to another may significantly change the level of forces available for deployment.

As Army forces continue to evolve based on operational deployments and worldwide situations, deployment demand must also evolve as an input to the rotational equipping model. The Army uses combinations of classified and unclassified scenarios for force planning, which can be easily integrated and checked with our unclassified assumptions above. For this analysis, we focused our equipping numbers on a fully rotational force. As this assumption for demand changes in the future as forces return from Iraq and Afghanistan and the Army finds a new steady state, these inputs will need to be updated.

About the Next Chapter

The next chapter shows the outputs of the rotational equipping model for various pieces of equipment, describes the equipment and units to which an RES would best be applied, and determines what the budget implications for some examples of that equipment would be if an RES were adopted. (The rotational equipping model that was built to take force structure, demands, and the rules of rotational equipping and determine how the rotational strategy would affect equipment inventories is described in the appendix.)

CHAPTER THREE

Applying the Rotational Equipping Strategy

An RES presents opportunities for the Army to reduce its total inventory of equipment without diminishing its ability to meet challenging deployment demands and conduct training. The RES pertains broadly across the Army's BCTs and support brigades, providing the greatest benefits when used to manage relatively expensive and abundant systems. In this chapter, we describe how the Army may apply an RES, compare the system numbers necessary under the RES to the current and planned inventory of several items, and highlight potential budgetary savings that the Army may gain when implementing an RES.

Applying the RES

The RES is designed so that units cycling through a readiness cycle, like ARFORGEN, receive an adequate level of equipment to meet training and operational needs during each phase of the cycle. The first step in implementing the RES is determining how and where it applies within the Army.

We examined Army organizations and the equipment assigned to them to determine the suitability of the RES. Conceptually, an RES is best applied when two separate conditions are met. First, there needs to be an adequate number of similar organizations in each readiness phase so that the equipment can rotate among them. And second, each unit requires an adequate stock of the equipment, which needs to grow as the unit progresses through its readiness cycle.

We examined Army organizations and their assigned equipment to identify the units and types of equipment where the RES may have the greatest impact in terms of reducing Army procurement and modernization investments. It is important to note that many types of equipment are used across many different unit types. For example, most Army organizations are assigned high-mobility multipurpose wheeled vehicles. When examining equipment like this, we included the entire set of units that use that piece of equipment.

Applicable Army Units for Rotational Equipping

We examined the entire Army force structure in order to identify how the Army may apply the RES. Some parts of the organization do not lend themselves to an RES because they do not follow a rotational readiness cycle; therefore, we eliminated the Army's theater-committed assets and generating force organizations from consideration.[1] The types of units to which the RES best applies are those that follow a rotational readiness cycle and have a great deal of equipment assigned; multiple units of the same type should also exist. The Army's different BCTs, multifunctional brigades, and some functional support brigades fit these criteria well. Table 3.1 highlights the units included in our RES model. All told, these units are well known to be rotational and constitute approximately 50 percent of the operating force. Therefore, as a starting for analysis, they are a reasonable place to look for efficiencies.

Appropriate Equipment Types for Rotational Equipping

Within the units mentioned above, not all equipment is amenable to an RES. The Army may apply the RES to most types of equipment within a given organization, with certain specific limitations. A limiting factor on the general application of the RES is the quantity of equipment that organizations need for individual and small-unit training in the early stages of the ARFORGEN cycle. Even though the RES is broadly

---

[1] Theater-committed forces include Army service component commands that are associated with a specific region and force stationed in Korea. The Army's generating force includes organizations that focus on training or sustaining the Army's deployable forces.

Table 3.1
Rotational Units

Unit Type	AC	RC	Total
Corps HQ	4	0	4
Division HQ	10	8	18
Infantry brigade combat team (IBCT)	20	20	40
Heavy brigade combat team (HBCT)	17	7	24
Stryker brigade combat team (SBCT)	8	1	9
Combat aviation brigade (CAB)	13	8	21
Fires brigade	7	7	14
Maneuver enhancement brigade (MEB)	3	18	21
Sustainment brigade	13	19	32
Battlefield surveillance brigade (BFSB)	3	7	10
Theater aviation brigade (TAB)	1	6	7

SOURCE: Data from Headquarters, Department of the Army, *FMSWeb (Force Management System Web Site)*, undated(b).

applicable, it provides the most benefit in terms of procurement cost savings when applied to expensive systems where there are plans for additional investment. We focus our analysis on these systems.

To identify the key systems in which the RES has the greatest potential for savings, we analyzed the equipment that the Army purchases. We limited our search to major end items in the Army that meet at least one of the following criteria: The Army is authorized to have a stock of at least 300, the total cost of the stock of equipment is greater than $100 million, or there is an identified shortage of stock valued at greater than $50 million.[2] There are 640 different pieces of equipment that meet one of the criteria for this search.

---

2 Using Army databases, we identified about 6,000 different pieces of equipment that the Army purchases. About 3,000 of these items are defined as major end items by the Army.

The search yielded very different types of equipment, including individual soldier items, vehicles, and weapon systems. The total stock of the different pieces of equipment authorized in Army units ranges from seven to over 900,000 items. The value of the stock of the different systems included in our search ranges from about $40,000 to over $14 billion.[3] Twenty percent of the items are ground vehicles, 10 percent are weapon systems, and all Army helicopters are included.

There are also items included that are useful to the Army National Guard (ARNG) for its state-controlled domestic role. Thirty-four percent of the systems we examined are identified by the Army as critical dual use (CDU) items, meaning that they are important to the ARNG for both "domestic and war fighting missions."[4] CDU items are important when considering implementing the RES, as these items should not rotate within the ARNG because of the need for these organizations to use this equipment to fulfill their domestic responsibilities. Since CDU equipment will not rotate through ARNG units, the potential for reducing the total stock of CDU equipment is less than for non-CDU equipment.[5]

After identifying candidate equipment to flow through the RES, we considered which items presented the greatest potential for savings by examining the total value of the equipment inventory, the number of rotational items, and planned future investments. Rotational items are those authorized to units identified in Table 3.1. We found that there are not any equipment types in which the entire stock is rotational, because some equipment is always assigned to units that do not follow a rotational readiness cycle, CDU items are not rotational in the ARNG, and we were not able to include the entire force in our rota-

---

3 We defined the total value of a type of equipment to be the number of systems authorized multiplied by the listed replacement cost.

4 Raymond W. Carpenter, "Statement Before HASC on Army and Air National Guard Equipment," Washington, D.C., May 5, 2009.

5 Statements from some Army organizations explained that CDU equipment was not appropriate for rotational equipping. We found that the reductions to the necessary inventory are less for CDU items, but cutting the inventory is still possible, especially when much of the force is in the U.S. Army Reserve and AC. This may also provide significant budgetary savings when large investments in a system are planned.

tional analysis. Figure 3.1 displays a set of items with large total inventory costs along with the stock of rotational items.

Equipment displayed in Figure 3.1 represents the types of equipment in which the Army has already invested at least $1 billion and includes helicopters, combat vehicles, other tracked and wheeled vehicles, and weapon systems. These systems are abundantly found in the rotational units listed in Table 3.1. For example, all of the helicopters displayed are found in the Army's CABs, tanks and Bradley variants are found in HBCTs, and trucks are found throughout rotational parts of the force. What this means is that the Army's most expensive and prominent systems to which a rotational strategy might apply actually

Figure 3.1
Candidate Equipment for the RES

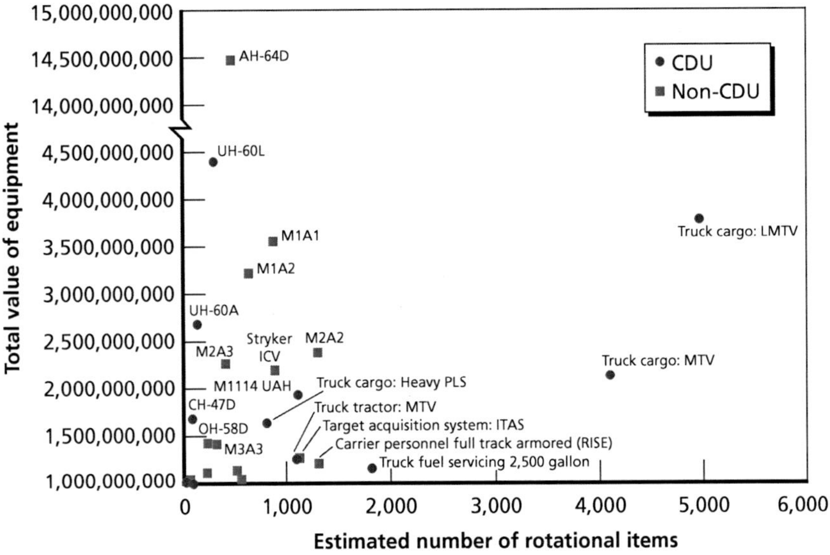

SOURCE: Data for figure collected from various Army sources and integrated by the study team.
NOTES: ICV = infantry carrier vehicle, ITAS = improved target acquisition system, LMTV = light medium tactical vehicle, MTV = medium tactical vehicle, PLS = palletized load system, RISE = reliability improvements for selected equipment, UAH = up-armored Humvee.
RAND *MG1092-3.1*

exist predominately in rotational units, meaning that an RES could have a potentially large impact on how the Army equips itself.

We also examined the procurement budget to identify where the Army plans on making large investments in the near future. In FY 2011, the Army procurement program accounted for $21.3 billion, nearly 15 percent of the Army's total budget. The budget includes several large investment programs in each of its four procurement categories: aircraft, weapons and tracked combat vehicles, missile systems, and other procurement programs. Procurement plans in these categories include programs for new equipment acquisition and the modernization of existing stock. The 15 largest active procurement investments through FY 2015 are in the systems listed in Table 3.2. Of these items, eight are candidates for the RES (with an additional three as potentially appropriate), totaling nearly $30 billion in Army expenditures over the Five Years Defense Plan (FYDP). A few would require additional analysis—particularly the subsystems, which may be able to rotate independently from their main platforms.

In some cases the Army budget also indicates total procurement expenditures and item quantities for the period beyond FYDP. Based on the FY 2011 budget, the most expensive procurement programs planned for the period beyond FYDP include the Black Hawk, Chinook, Apache, Paladin, and FMTV. Each of these programs has planned beyond-FYDP expenditures of at least $3.5 billion.

Based on our analysis of Army forces and the equipment assigned to them, we focused our analysis of the effects of an RES on a select number of systems from FY 2012 on. These systems are the AH-64 Apache and CH-47 Chinook helicopters, as well as the M109 Paladin self-propelled howitzer. Both the Apache and Chinook procurement programs include sizable new purchase and remanufacturing components. The Paladin program exclusively modernizes existing stock. The characteristics of each of the three programs are shown in Table 3.3.

Table 3.2
Major Item Procurement Program Expenditures for FYs 2011–2015

System	Expense (millions of dollars)	Quantity	Consistent with RES?
UH-60	7,271.3	378	Y
CH-47	5,531.4	195	Y
AH-64	5,305.9	219	Y
WIN-T	5,070.7		Maybe
FMTV	3,107.2	8,726	Y
JTRS	2,718.0		Maybe
MQ-1 system	2,468.8	134	Y
Counterfire radars	2,402.5	174	Y
I3MP	1,899.2		N/A
FHTV	1,746.0	26,761	Y
GMLRS	1,617.5	14,112	N
DCGS-A ISR	1,546.7		N
Nonsystem training	1,546.4		N
Paladin	1,395.1		Y
SIIRCM	1,322.5	222	Maybe

SOURCE: Data from Fiscal Year 2011 Committee Staff Procurement Backup Book Budget Estimates, at Army Financial Management—Assistant Secretary of the Army for Financial Management and Comptroller, *Budget Materials—Fiscal Year 2011*, last modified November 11, 2010.

NOTES: DCGS-A ISR = distributed common ground system—Army, intelligence, surveillance, and reconnaissance; FHTV = family of heavy tactical vehicles; FMTV = family of medium tactical vehicles; GMLRS = guided multiple launch rocket system; JTRS = joint tactical radio system; SIIRCM = Suite of Integrated Infrared Countermeasures; WIN-T = Warfighter Information Network—Tactical. Some budget submissions omit quantities.

Table 3.3
Procurement and Expenditures for Apaches, Chinooks, and Paladins

System	Procurement Quantity	Expenditure (millions of dollars)
AH-64 Apache Block III total	666+	11,418
Remanufacture	610	8,020
New purchase	56	2,151
Other modifications		1,246
CH-47 Chinook total	225+	14,168
SLEP/ReNew D to F	161	4,458
F, new purchase	64	2,073
Other modifications		7,637
Paladin howitzer upgrades	525	2,911

SOURCE: Data from the Fiscal Year 2011 Committee Staff Procurement Backup Book Budget Estimates.

NOTES: The quantity of items affected is not available for other modifications to AH-64 and CH-47. Because of rounding, some amounts may not total correctly.

Application of Rotational Equipping to Select Equipment and Units

Using the Rotational Equipping and Modernization Model (REMM) detailed in the appendix, we determined the size of the fleet needed to meet theater demands for a 20-BCT steady-state scenario.[6] We computed two different equipping policy scenarios, which we believe illustrate a range of risk in a rotationally equipped force. The top of the range we have termed the *rotational-high level* and the bottom of the range the *rotational-low level.*

[6] Recall that the 20-BCT steady-state scenario includes a 40-BCT one-year surge and runs for 15 years.

In both cases, the tables of distribution and allowances (TDA) units are always equipped to 100 percent of their required amount of equipment for centralized training and testing. This ensures that the training base is not eroded throughout our simulated scenarios. Similarly, the Army prepositioned stock (APS) equipment is always filled to 100 percent, as preparation for overseas contingencies. We have kept these numbers stable, with the possibility of changing the numbers as policies are updated. Both cases take into consideration CDU items, such as CH-47 cargo helicopters or trucks, which must be filled to 100 percent at all times for ARNG units possessing these items. This is based on the current policies governing what equipment is CDU. The main differences between the rotational-high and rotational-low levels are explained below.

The Rotational-High Level

The rotational-high level of equipping is the relatively low-risk strategy for equipping a rotational force driven by reductions in the amount of equipment that units have during early phases of ARFORGEN. We assume that units in the Available phase require 100 percent of the authorized MTOE equipment, units in the Ready phase starting one year before deployment require 90 percent of authorized MTOE, units in collective training 200 days before the Ready phase require 85 percent of authorized MTOE, and units in individual training require 50 percent of authorized MTOE. During a period of national emergency, when demand surges from 20 BCTs to 40 BCTs, units remaining in CONUS in the Reset phase keep their complement of equipment for training, be it 50 or 85 percent. The levels presented here are comparable to those presented in the 2009 Army Equipping Strategy.[7]

The Rotational-Low Level

The rotational-low level is a relatively higher-risk equipping strategy driven by the minimum amount of equipment that the Army needs to have somewhere in its force to meet surge demands. We assume that units in the Available phase require 100 percent of the MTOE equip-

---

[7] Headquarters, Department of the Army, 2009.

ment, units in the Ready phase one year before deployment require 75 percent of MTOE, units in collective training 200 days before the Ready phase require 50 percent of MTOE, and units in individual training require 0 percent of MTOE and will receive training centrally instead of at unit locations. While centralized training may incur extra items beyond those currently existing in TDA training units, the exact number is left to further work.

We note that for AC units, the amount of time at 0 percent MTOE is about 15 weeks immediately after deployment. This corresponds to soldier travel and unit reconstitution time, when little training will likely take place. During a period of national emergency, when demand surges from 20 BCTs to 40 BCTs, we discard the requirement that units remaining in CONUS in the Reset phase possess their complement of equipment for training and instead allow for units deploying to theater to pull equipment from units remaining in CONUS, reducing the overall requirement for items.

Application to Various Items

The rotational-high and rotational-low equipping policies therefore create a band of total equipment numbers from which a policymaker can choose when considering implementing a rotationally equipped force for a given type of equipment in a given type of unit. Table 3.4 illustrates the rotational-high and rotational-low equipping requirements for a variety of major end items, as calculated by our model.[8]

As shown in the table, the major end items listed include prominent Army systems in which considerable investments are made each year to upgrade, modernize, and reset from prior use. If the RES were to be employed with the parameters described previously, the total inventory would be reduced up to approximately 25 percent, with some inventory reductions having much smaller impacts because of the nature of the force structure and relationship to the type of equipment.

---

[8] Note that for aviation units, we assumed a combat aviation brigade force structure of 13 (based on the 2010 QDR [U.S. Department of Defense, 2010]) and an additional nine HH-60 Medevac companies in the ARNG and Army RC. The CH-47, UH-60, and HH-60 were treated as CDU items of which possessing ARNG and RC force units required 100 percent on hand.

Table 3.4
Rotational-High and Rotational-Low Numbers for Some Major End Items

System	Total MTOE + TDA + APS	Rotational-Low Number (% of Total)	Rotational-High Number (% of Total)
AH-64D	772	550 (71%)	688 (89%)
CH-47	404	368 (91%)	398 (98%)
UH-60	1,467	1,351 (92%)	1,446 (98%)
HH-60	570	522 (92%)	561 (98%)
M109	616	454 (73%)	567 (92%)
M1A1	1,100	878 (80%)	1,028 (93%)
M1A2	832	664 (80%)	774 (93%)
M2A2	998	753 (75%)	905 (91%)
M2A3	953	778 (82%)	889 (93%)
M2A2ODS	351	246 (70%)	316 (90%)
M7 BFIST	315	227 (72%)	276 (88%)
M3A2	564	450 (80%)	520 (92%)
M3A3	402	314 (78%)	368 (92%)

NOTE: The data above only include CABs, theater aviation brigades, fires brigades, and HBCTs for all equipment.

Budgetary Effects of Implementing the RES

With an RES, the Army can reduce its procurement requirements by buying only enough materiel to equip its organizations based on their phase in the readiness cycle. Modeling an RES through a challenging, simulated deployment cycle ensures that the Army procures enough equipment so that it is able to surge deployed units by hastening movement through the readiness cycle. There are several billion dollars of potential savings over the next five years if the Army fully implements an RES. This would allow the Army to invest in other priorities without diminishing the capabilities provided by its equipment base.

In order to explore those impacts, we have chosen a few specific examples of equipment to link to current and proposed budgets: the Apache, the Chinook, and the Paladin. These cases represent high-priority equipment to which an RES would be beneficial.

In implementing an RES, decisions must be made as to when and where to make reductions in future acquisitions plans. The Apache helicopter with the current Apache Block III (AB3) upgrade program provides an example of different ways that the Army may choose to reduce the size of modernization efforts to achieve equipping levels implied by the RES. There are currently 682 Apache helicopters in the Army, and the AB3 program includes purchasing 56 new aircraft in FYs 2013–2015 at a cost of about $2.15 billion and remanufacturing 635 previous-generation Apaches.[9]

The Army's current Apache plans will produce 141 more helicopters than necessary using the rotational-low levels or three more than rotational-high levels. We examined two alternatives to reach either the rotational-low level of 550 AB3 helicopters or the rotational-high level of 688. The first alternative makes all cuts at the end of the AB3 program, when Block II (AB2) aircraft are being upgraded to the AB3 configuration. Under this alternative, the program will complete by 2021 and save about $1.5 billion equipping to rotational-low.[10] Based on the rotational-high target, the Army will save approximately $30 billion nominal, forgoing three upgrades from AB2 to AB3 in the out-years.

Alternatively, the Army may choose to cut the new production aircraft earlier in the program. Since new builds are much more expensive than remanufacturing existing helicopters, the savings from equipping to the rotational-low level are estimated to be twice as large and are earned earlier (about $2.1 billion in new builds avoided in FYs 2012–2015 and $0.9 billion in upgrades avoided in FYs 2023–2024) than the savings from the scenario preserving new builds. Under this alternative,

---

[9] This occurs from FY 2010 to FY 2024 and is at a significantly lower cost per unit than the new-build aircraft. Sources: Defense Acquisition Management Information Retrieval, *Selected Acquisition Report: AB3*, Washington, D.C.: U.S. Department of Defense, December 31, 2009; and Army Financial Management—Assistant Secretary of the Army for Financial Management and Comptroller, 2010.

[10] In this situation, the Army may retain a surplus stock of 188 AB2 helicopters.

the program will complete in 2023. The savings under rotational-high are the equivalent of forgoing three new-build aircraft, saving approximately $115 million nominal compared with current plans. Figure 3.2 highlights our estimates for the evolution of the Apache fleets given these two alternatives. The figure shows the changes over time in distribution among the two AH-64D blocks (AB2 and AB3) and AH-64A models.

The present plan for CH-47F Chinook acquisition and modernization, including the upgrade (ReNew) program and new-build purchases, began in FY 2003 and is scheduled to end during FY 2017. Approximately 225 of the planned 523 aircraft remain to be procured, totaling roughly $6.8 billion nominal in expenses yet to be realized from FY 2012 on. By equipping to the rotational-low level, the total CH-47F need may be reduced by as many as 74 units. The rotational-high level would result in a 44-unit reduction. In the rotational-low case, the similarity in costs under the ReNew program and new-build purchases over the course of the entire remaining program results in equivalent estimated savings under either of the two alternatives out-

Figure 3.2
Apache Inventories over Time Under Different Equipping Alternatives

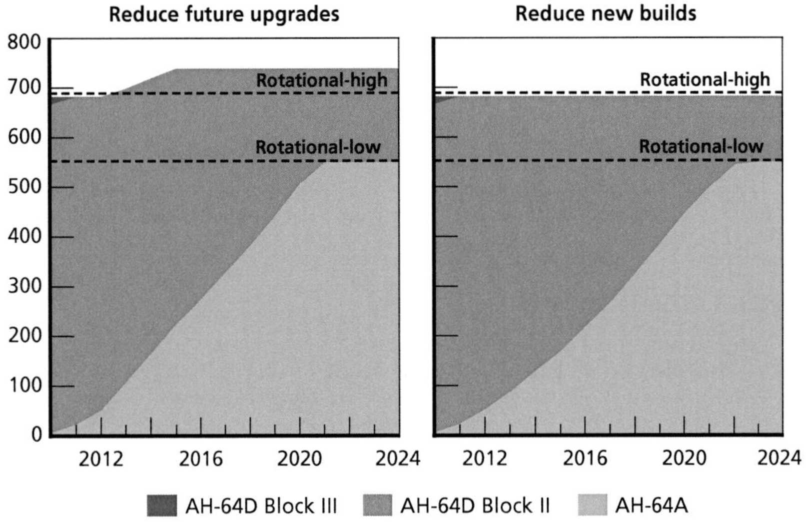

RAND *MG1092-3.2*

lined in the previous Apache example; the estimated budgetary reductions are on the order of $2.3 billion over FYs 2015–2017 when directing all cuts to the end of the program or over FYs 2012–2017 when targeting the more expensive new-build program first. Given the relatively higher expense of new builds early in the production cycle compared with the costs of the ReNew program, equipping to rotational-low by cutting the more expensive new builds first will save an estimated $1.5 billion versus $1.4 billion by cutting builds from the end of the schedule, regardless of cost.

For systems with a single associated procurement program, the analysis is more straightforward. The M109 Paladin modernization program does not have an associated new purchase program, thus the savings seen in equipping under RES rather than the current planned level are purely driven by unit cost over the time horizon selected by Army leadership. For illustrative purposes, the current upgrade program for the Paladin includes 543 platforms, fewer than the 567 needed to fulfill rotational-high levels. Equipping to the rotational-low level requires 89 fewer Paladin upgrades than current plans, resulting in an estimated $0.5 billion nominal savings in the post-FYDP period.

The impacts on the budget can be further expanded to other materiel. Each of the major pieces of equipment we discussed in this report has significant modernization efforts in the current five-year defense spending plan. Similarly, an RES might affect some of the Army's longer-term modernization efforts, such as those for the ground combat vehicle and joint light tactical vehicle. As those move into production and acquisition objectives are finalized, the total number procured might feasibly be reduced.

Conclusions About the Budget

The RES provides opportunities to reduce the total amounts of some equipment in the force. The equipment most amenable to an RES is typically the most expensive, is available in sufficient numbers for swapping among units, can easily be integrated from unit to unit, and is not limited by significant training needs. These attributes describe many of the Army's major end items.

The effect that smaller inventories have on the budget is threefold. First, with a smaller fleet, the Army can **reduce the number of systems needing significant repair or recapitalization** in the near term. For example, those tanks, Bradleys, and tactical wheeled vehicles most in need of repair could be divested in lieu of repair until the RES level of equipping is met. The actual savings of this strategy must still be determined but would have to be weighed against any increase in maintenance costs that a small, but more heavily employed, fleet might incur.[11]

Second, there are potential medium- and longer-term savings from **reducing the total number of systems needing upgrades**. The Army's modernization programs rely on upgrades to major systems, such as aircraft and tracked vehicles. If a rotational strategy is adopted that allows the Army to lower the overall number of systems needed, those upgrade programs could reduce the tail ends of the programs to save money. (Note that reducing near-term upgrades is also a possibility but would sacrifice some near-term capability.) As an example, the CH-47, UH-60M, and AH-64 upgrade programs total over $30 billion from FY 2012 and beyond. Finding a means of reducing the total stock necessary to upgrade will have profound impacts in the mid- to long-term budget, and our analysis shows that single-digit billions of dollars of savings are likely, given a commitment to any level between rotational-high and rotational-low. Based on the analysis shown in this report, we estimate that reducing inventories for the Paladin, CH-47, and AH-64 to rotational-low levels of equipment would avoid between $1.7 billion and $4.0 billion in upgrade costs for these major systems over the life span of the current modernization programs. This depends on whether new builds are forgone or not, in addition to reduction of current inventories. Alternatively, reducing inventories to rotational-high levels may save between zero and $1.1 billion over the same period.

---

[11] Analysis has not yet been performed to determine which case costs more for specific platforms—smaller fleets used intensely or larger fleets used sporadically.

Lastly, a smaller inventory for major rotational items can provide options for **reducing or even eliminating new purchases in the near term**. Some new force structure and replacement aircraft and other vehicles and weapons are predicated on an MTOE-based equipping strategy. A reduced number may offer the opportunity to eliminate those purchases. As an example, the two new CABs that the Army is standing up based on the 2010 QDR will lead to new purchases of both CH-47s and AH-64s, whose new purchases from FY 2012 onward total approximately $4.5 billion combined. To find ways of adapting those fleets to lower overall numbers without sacrificing training demands or deployment needs could have a large, near-term impact on the budget.

CHAPTER FOUR

Conclusions

The Army's ARFORGEN-based approach to providing forces to meet national needs is well under way. Since the early 2000s, the Army has been generating units to fight two major wars in Iraq and Afghanistan, which has put stress on the force and highlighted the costs of equipping and modernizing a force currently in high use. While many facets of a rotational strategy have been implemented, an RES is only beginning to be considered and implemented.

In this project we further detailed the important attributes of an RES and how it might work within the Army's ARFORGEN model. The analytical model we developed showed how the Army's force structure and rules of ARFORGEN deployments can be used to calculate the smaller inventories necessary to meet rotational needs. With those new inventories tailored to a force focused on rotational deployments, the Army might have numerous opportunities to reduce investments in future budgets and bring the equipping strategy more in line with personnel policies.

The analysis in this report entailed a detailed modeling effort and thus rests on a number of assumptions that will need to be understood, monitored, and updated as circumstances warrant. The analysis is highly dependent on a few of those assumptions, notably the demand assumptions and rules of how ARFORGEN is executed (see the appendix for additional information on how demand and other inputs drive the outputs of the model). Other assumptions about rates of equipment lost in combat and transit times are also important and thus amenable to additional analysis and strategies to reduce or control risks.

Nonetheless, careful consideration of the assumptions underpinning the analysis can offer insights into how an RES would affect equipping and modernizing the Army.

Policy Decisions

The assumptions described above, therefore, can be considered policy decisions that the Army faces, should it fully execute an RES. There are three assumptions described in this report that bear repeating:

1. The Army will have to decide what demand it will assume for planning purposes or whether its force structure and rules of ARFORGEN will drive that number.
2. The Army will have to decide how much equipment each unit will get for each piece of equipment during the early phases of ARFORGEN.
3. The rules of ARFORGEN—particularly those associated with deployment and dwell times—need to be formally adopted for equipment planning purposes.

In this report, we assumed answers for all of these decisions through rationales built from senior decisionmakers within the Army, such as the Army's Chief of Staff, and found considerable cost savings potentially available.[1]

Implications

An RES would reduce total equipment and may force a different process of equipping rotational units. By reducing fills in early ARFORGEN phases, the Army can reduce total stock necessary for some major end items. Also, by structuring the equipping rules for periods of surge, the Army might find additional reductions in stock, but this needs to

[1] The appendix of this report includes numerous deviations from those assumptions as context for future Army policy decisions.

be weighed against other risks with regard to shortages in protracted, high-intensity environments.

The equipping strategy can be executed in such a way that costs can be saved without sacrificing capacity and capability. Our analysis showed how the structuring of rules, predicated on changes to the way units report equipment readiness, can allow the Army to meet its steady-state and surge demands for equipped units while at the same time saving costs in near- and far-term Army budgets. Reducing equipment affects budgets by reducing new equipment purchases on a path to lower equipment levels, reducing upgrades as fleets are upgraded to a lower acquisition objective, and divesting of considerable equipment, as well as ensuring that those costs are carried through all portions of the budget. The RES will entail careful management, however, as equipment must constantly move among rotational units.

Our preceding analysis also highlighted some limitations in applying the RES. The strategy is not a wholesale selloff of Army assets but rather should be applied to certain types of equipment and units. The broad criteria for determining high-impact areas in which such a strategy might best be applied are detailed here in this report.

Additionally, and perhaps most importantly, an RES is a significant change in how the Army equips units and therefore a potentially significant change in how soldiers view units and the Army. While this analysis does not detail the cultural changes necessary to implement such a rotational strategy, it will be necessary to include such considerations as the Army moves forward.

APPENDIX

Modeling the Rotational Equipping Strategy

Introduction

In order to evaluate the impacts that the RES might have on equipping and modernization, we built an equipping model based on force structure, planning factors for demand, and rules and policies for equipping a rotational force. The RES, as described by the Army and which we model for this project, is a significant departure from the Army's current practices.

Currently, unit commanders are required to report their readiness levels by determining what percentage of their total requirement they have on hand and capable, regardless of how far they are from deployment in the ARFORGEN cycle or how much equipment their unit requires for training. With little incentive for relinquishing equipment, Army operational forces typically take and keep possession of their full set of equipment from training through deployment. In our model, we relax equipment requirements in early phases of ARFORGEN while preserving the combat requirement for units deploying to theater.

Overview of Our Model

The goal of the modeling effort was to illustrate in concrete, quantitative terms the effects of strategic choices in equipping a rotational Army. The model simulates the flow of equipment and personnel on a day-by-day basis to meet deployments and training needs in order to determine how much equipment the rotational force ultimately requires.

The Rotational Equipping and Modernization Model (REMM) is divided into three successive parts.[1] The Demand portion uses planning factors for units and unit MTOE lists to derive the demand for individual major end items over time. The Supply portion creates a virtual pool of items and stochastically invokes a loss and depot model to determine how many items are available and are at each location over time. The Transportation portion then optimizes to determine the transport costs required to move items to fill demands. The fidelity of the simulation not only is sufficient to examine the "top-line" number of items that the Army would require to implement any given equipping strategy but also is useful to show the level of operational risk of that strategy. Thus, as equipment levels are lowered within the Army, the model can illustrate which units have to "go without" in order to meet theater demands.

Various Assumptions Are Necessary

The model makes a number of assumptions. As mentioned in Chapter Two, the planning factors for demand are a very sensitive input to what equipment the Army will ultimately need to stock. The planning factors we use for demand are based on the guidance explained previously and limit the fraction of the total Army that will be deployed at any given time. In other words, in no scenario is every brigade in the Army simultaneously deployed in combat.

We also assume that during periods of necessity, such as multiple simultaneous major combat operations (MCOs), unit demands for equipment can be prioritized to units later in their rotational cycle, allowing units in the Ready phase to deploy prematurely by transfer-

[1] The model we developed is written in Java and is typically natively compiled using just-in-time compilation on the target platform. Identical versions can be compiled for Windows, OS X, and Linux, allowing for maximum flexibility for deployment. Further information on the details of the REMM will be produced in a subsequent technical report.

ring equipment from units in the Reset phase. Similarly, equipment can be in motion to meet the needs of the moving units.

For the purposes of the analysis shown here, we have elected to keep all TDA units at 100 percent of the authorized equipment at all times in order to meet training and test missions. Some of the cases we run do test how raiding or preserving the TDA equipment during large-scale deployments affects the ultimate amount of equipment that the Army needs to keep in inventory.

We also assume many factors for use in the modeling. Equipment is destroyed in combat or lost through training accidents at a rate that can be estimated for a given theater or CONUS training regimen. These numbers were generated from current estimates in theater and training and can be updated. Similarly, times in depot are estimated from current practices.

Force Structure Assumptions

The results in this document are derived for systems residing in combat aviation brigades (CABs), theater aviation brigades (TABs), heavy brigade combat teams (HBCTs), and field artillery brigades (FA brigades). The following assumptions were made regarding MTOE force structure.

CABs: A total of 21 CABs are included in the model, with 13 in the AC and eight in the ARNG. Eight of the CABs are CAB-heavy, with two attack/recon battalions of AH-64s. Six are CAB-medium, with one attack/recon battalion of AH-64s. One is a CAB-light, and six ARNG CABs are CAB-expeditionary. Each CAB has a full combat requirement of 12 CH-47s, 15 HH-60s, 38 UH-60s, and either 0, 24, or 48 AH-64s (for light, expeditionary/medium, and heavy CABs, respectively).

TABs: Four ARNG TABs are included in the model. Two TABs are equipped with 36 CH-47s, 45 HH-60s, and 24 UH-60s. The other two TABs are equipped with 12 CH-47s, 15 HH-60s, and 98 UH-60s. These TABs are supplemented with nine Medevac battalions in the Army RC and ARNG with 15 HH-60s each.

HBCTs: Twenty-four HBCTs are used in the model. Each HBCT has 58 M1A1 or M1A2, 61 M2A2 or M2A3, 13 M2A2ODS, 11 BFIST, 29 M3A2 or M3A3, and 16 M109.

FA brigades: Nine FA brigades are used in the model, with 12 M109s each.

Inputs to the Model

The REMM is an exploratory analysis tool to enable the calculation of the effects of policy decisions on equipment stocks. This assists the user in quickly adjusting and exploring the effects and sensitivity of various inputs to the total equipping requirements. The data requirements are easily input by a user for exploration and fall into three main categories: demand parameters, force structure parameters, and transportation parameters. Each is explained below.

Supply Inputs

The model currently runs at the battalion-and-above level. The model uses a list of units (both MTOE and TDA units), their affiliations with other units (e.g., who tends to deploy with whom), and an associated list of major end items under analysis. Multiple items in each unit can be assessed simultaneously. Each unit's location is also needed, along with transport time between locations. The force structure is easily built from the Army's current structure or from plans for the future, depending on the scope of the analysis. In all cases here, we used unclassified force structures based on near-term "to-be" forces from the QDR and other Army documentation.

Demand Inputs

REMM uses demand curves over time for each unit (or group of units) in a deployable location. These are calculated in a separate effort to reflect the force structure–based demands. In many cases, the demand for units is based on affiliations with other units, which can be built from historical experience, subject matter expert (SME) input, or Army plans for the future. In all cases, these inputs are tailored to Army policies and are left as inputs to the model.

Rule Inputs

The model's internal structure, which defines the relationships between the supply and demand of forces, is governed by interpretations of the RES. Each input is built from prior examples, specific guidance, or SME input. Inputs include starting and ending days for Reset, Training, Ready, and Available phases and subphases for each force component (AC, RC, and ARNG); fill rates of equipment as a percentage of MTOE for each phase and subphase; and prioritization of units for equipment fills and deployment.

There are also various rules about the equipment—e.g., how often the equipment is sent to depot or on-installation facilities for maintenance reset or recapitalization, is destroyed, or is upgraded. These inputs are built from discussions with SMEs and current practices and help to highlight additional areas where policies might change effectiveness.

Outputs

REMM internally simulates the location, phase, and deployment status of every brigade and battalion, provided as force structure inputs, day by day over a multiyear period. It also tracks the demands for equipment by those units and simulates and tracks the movement of individual items of equipment that meet those demands. The model produces numerous outputs to help policy analysis:

1. phase and locations on a day-by-day basis for every brigade and every unit in that brigade
2. demand for items by location and by unit on a day-by-day basis consistent with fill rates
3. aggregate demand across the Army for items on a day-by-day basis across the entire force
4. met and unmet demands for items at each unit and location

5. locations, assignments, and status (deployed, in transit, in depot, lost/destroyed, etc.) for each individual tracked item on a day-by-day basis
6. maximum and minimum demands for items consistent with input policy levers and MTOE fill rates
7. BOG:dwell ratios for individual units and across force components
8. equipment and depot utilization rates
9. time spent in transport for individual items and across a fleet of items
10. individual unit-to-unit item transactions.

Demand

Army equipping strategy documents and previous studies considering a rotational force have typically assumed that one-third of the active force and one-fifth to one-sixth of the reserve force is either deployed or in the Available stage of the ARFORGEN cycle at any given point in time. While this time-averaged estimate is useful as a starting construct for planning factors, it is insufficient to describe the ebbs and flows of aggregate demands that occur throughout the Army as a limited number of brigades respond to demand signals. As an example, consider the planned 13 AC and eight ARNG CABs. At any given time, if we consider only the gross planning factors, we could assume anywhere from five AC CABs and one ARNG CAB deployed to six AC and two ARNG CABs deployed in steady state, an effective difference of hundreds of helicopters needed.

These planning factors also cannot anticipate how the deployment response of Army units to an arbitrary demand signal will change the need for equipping. For example, following the return home of units responding to a surge in theater demand for deployed units, the ratio of units in the Reset phase will be large compared to those in the Ready/Train phase. If units in the Reset phase have lower equipping needs than those in subsequent training phases, we would overestimate the total equipping needs following major combat operations if we assumed that units immediately returned to a cycle in which they were uniformly distributed among the steady-state rotation phases.

Implementation

REMM's demand calculations demonstrate how a given force structure will respond to theater demands for units, consistent with Army policy and empirical evidence on how demand signals have been met by the rotational force over the last several years in Operation Iraqi Freedom (OIF) and Operation Enduring Freedom (OEF). The model begins by assigning every brigade type under consideration (such as the whole set of HBCTs or the whole set of CABs) a "phasing" in their rotational cycle. For AC brigades, this cycle is three years long, and for RC and ARNG brigades, the cycle is either five or six years long. The phasing is a day in the three- or five-/six-year cycle, respectively, chosen to balance the steady-state supply of brigades evenly across the simulation timeline of at least 15 years. It is assumed that all units subordinate to a brigade rotate with their brigade; i.e., no battalions are detached and deployed with other brigades, although this is in practice known to happen despite the doctrinal desire for Army modularization based on full brigade deployments.

These initial unit phasings are initial conditions that then respond to a set of user-defined demand signals over the length of the simulation for the number of each type of brigade. If the demand signal requires more brigades deployed than are available at a given timestep from the initial phased rotation cycles, the model will "break" the cycle of the unit closest to deployment and pull that unit to theater. Likewise, if the demand for brigades in theater decreases below the number of brigades deployed, the model can send units home early and will balance the phasing of all units in the system to compensate, extending reset and ready phases as necessary to maintain a constant supply of brigades for steady-state operations while meeting rest and training time requirements. See Figure A.1.

The result of the initial phasing of unit rotations and subsequent adjustment of phasings to theater demands is the rotational status (Reset, Ready, and Available phases and any user-defined subphases

Figure A.1
REMM Demand Model Overview

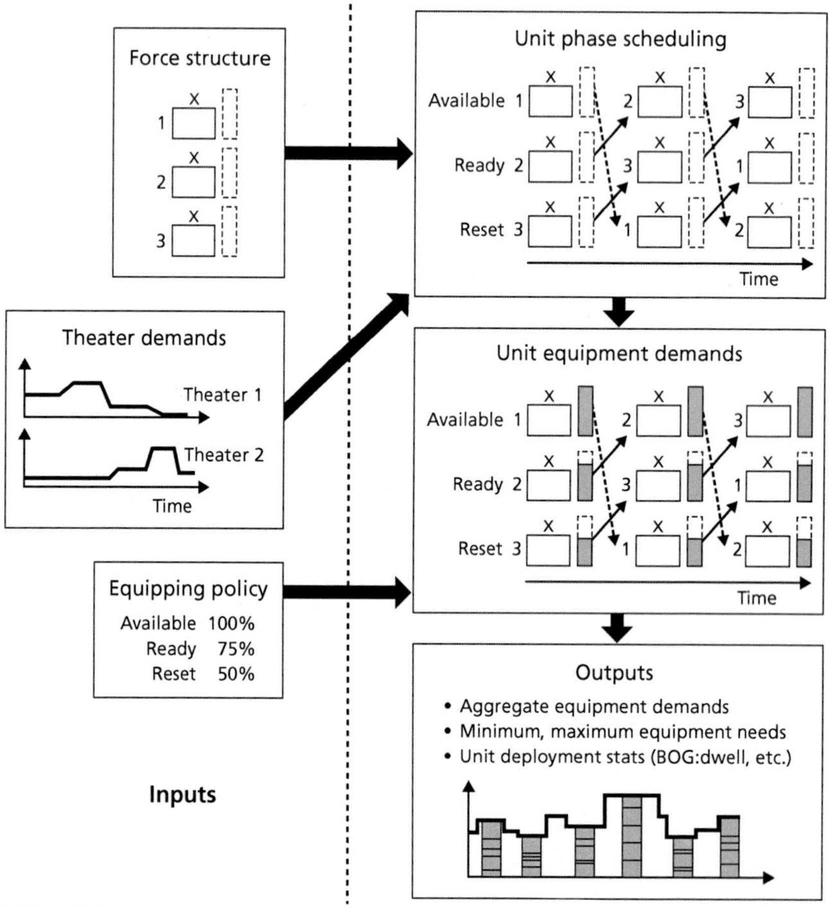

RAND *MG1092-A.1*

thereof) of every brigade, and therefore every battalion, by timestep. These are combined with percentage of MTOE requirements for each item for each phase to determine the unit-level requirements for items at each timestep, and these can be integrated across the Army to determine the aggregate demand for items across the simulation length.

The user can define location information, including unit home stations and the theater in which demands occur. A derivative result of the demand calculation is therefore also a list of demands by timestep at each location defined in the user input. This will become relevant when considering the transport cost model.

The peak of the aggregate demand curve for a given class of equipment is the most basic result of the model. It determines, absent of items to compensate for depot repair, combat losses, training accidents, and items in transport, the actual number of items that the Army needs on its "worst day" throughout the simulation for a given demand signal and force structure to satisfy both its theater demands and its training requirements. Note that if we set the procurement objective to this "worst day" number, units will often have more items on hand than the minimum percentage of MTOE defined by the user throughout the simulation cycle, with the exception of that day.

M109 Example

As a concrete example, consider the M109 Paladin self-propelled howitzer. In our model, we include 36 fires battalions distributed among 18 HBCTs (12 AC and six ARNG) and nine fires brigades (six AC and three ARNG). Each HBCT is assumed to be interchangeable for the purposes of demand with all other HBCTs, and likewise each fires brigade is interchangeable with every other fires brigade. The REMM demand model begins by choosing cycle start dates for each brigade over the user-specified simulation length, aiming to keep about five HBCTs and three fires brigades in the available phase with as little variation as possible. See Figure A.2.

We then input a stressing theater demand curve over 15 years that consists of a seven-HBCT steady state and a 14-HBCT surge, as well as a steady state of three fires brigades and a surge of six fires brigades. The surge lasts for exactly one year and occurs in the middle of the 15-year simulation period. REMM considers each day on the theater demand curve without regard to the future; it breaks the rotational cycle for each type of brigade and sends its units to theater as necessary.

If we assume that for the first 100 days of the rotation an M109 battalion needs no equipment while it recovers and reconstitutes from deployment, for the last 200 days of its time in the Reset phase it requires 50 percent of MTOE (either eight or nine M109s per battalion), in the one-year Train/Ready phase it requires 75 percent of

Figure A.2
Simulated Demand for M109 over 15 Years

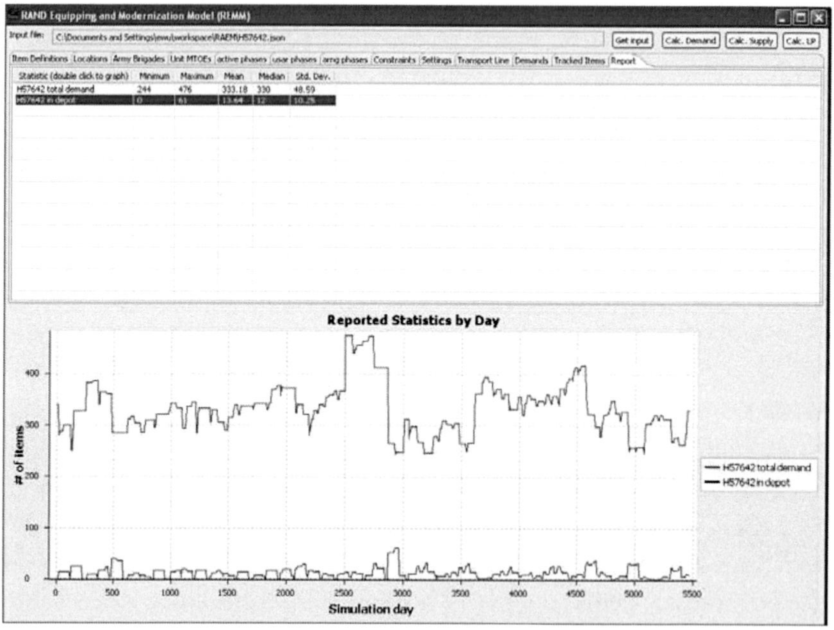

NOTE: The yearlong increased demand for equipment at day 2,500 is driven by a notional multi-MCO scenario requiring a total of 40 BCTs deployed.

RAND *MG1092-A.2*

MTOE (either 12 or 14 M109s per battalion), and while in the Available/Deployed phase it requires 100 percent of MTOE, we then derive the results that during the one-year surge we require a total of 476 M109s across these units in order to satisfy at-home training requirements, as well as fully equipping deployed units.

Supply Model

Determining the equipment demands as a result of theater deployment demand curves is insufficient to understand the full stress that an equipping policy of equipping units early in their rotational cycle under full MTOE combat-level requirements entails. If the Army procures or divests equipment under full MTOE equipping across the force, risk may be incurred during periods of high stress on the Army, such as simultaneous MCOs or sustained, multiyear irregular warfare deployments. In particular, because of the need for equipment to be sent to depot maintenance after deployment to theater, combat and accident losses, and transport time, simply evaluating whether there are enough pieces of equipment to draw from units in the Reset and Training phases to top off prematurely deploying units is not sufficient to evaluate risk.

The purpose of the supply model is to stochastically simulate probabilistic events that may occur when considering an equipping strategy with constant motion. By doing so, we can gauge how theater demands; force structure; and the repair, replacement, and transport properties of a piece of equipment can affect the stresses on an equipment fleet as a result of equipping policy.

Implementation

After the demand model runs and reports to the user the appropriate maximum number of items required throughout the simulation cycle, the user can choose to run a supply model with this number or set a number of his or her own choosing. This overall fleet-size number is the primary input.

The supply model takes as its secondary inputs the loss rate of an item in each phase (typically this number is much higher for combat vehicles, such as the M1 or the AH-64, in theater than it is for noncombat vehicles in CONUS), the number of days (typically at least two years) required to procure a replacement for a total loss, the probability that an item must be sent to depot maintenance following the conclusion of the time the unit holding it is in theater, and the number of days an item must remain in depot maintenance in order to be rotated back into the general pool. See Figure A.3.

When invoked, the supply model creates a set of *N* virtual items, where *N* is either the maximum demand determined in the first calculation or the user-defined value. On a timestep-by-timestep basis, the combat/accident loss model and depot model determine the number of items unavailable for assignment, *U*. The demands for individual

Figure A.3
REMM Supply Model Overview

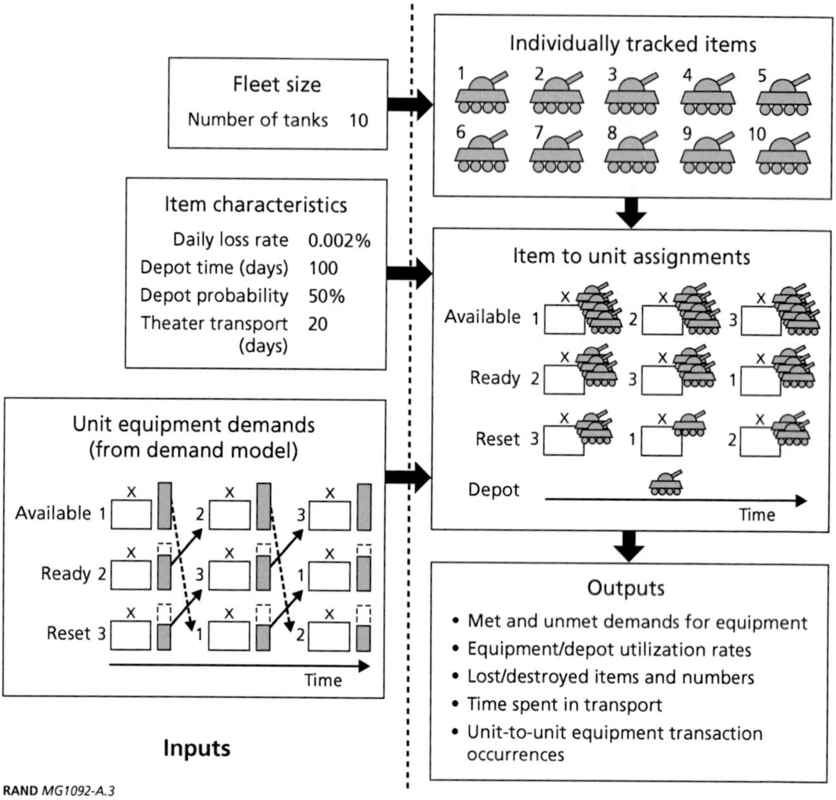

RAND *MG1092-A.3*

items by each unit are sorted at each timestep in order of priority, where units in theater receive the highest priority, followed by units in the Ready phase, and finally units in the Reset phase; degeneracy within the phases is broken by prioritizing units that are later in their rotation cycle.

Finally, the $N - U$ available items are distributed among the units in the prioritized order of their demands, without regard to transport costs. These available items are assumed to be identical, and the task of sorting which item (identified by a virtual serial number) is at which unit and location is determined by the transport submodel.

The transport submodel currently models only item transport times to and from theater. We do not yet include CONUS transportation costs, as the number of trips per item should be roughly the same but will be phased differently. For example, instead of receiving all of its M1 tanks at once during the Reset phase, a unit may receive its set of tanks incrementally as it progresses through the Reset and Ready phases, resulting in an equivalent number of trips with respect to current policy. We focus instead on modeling theater transportation times, which are particularly important in surge scenarios where deploying units may draw equipment from APS, units in Reset, or TDA units.

Assumptions

Various assumptions were made for many parameters in the model:

- **Transport time to and from theater.** We assumed that equipment would take anywhere from 20 to 60 days to deploy to theater from CONUS, corresponding to either air or sea movements.
- **Equipment attrition.** Using publicly released data on helicopter attrition incidents and requests for replacements, we derived equipment attrition rates that are used stochastically in the model to remove items from the supply model. In theater, we assume a daily loss chance per item of 0.001 percent, while in CONUS for training, we assume a daily loss chance of 0.002 percent per item.
- **Increased needs for maintenance.** The Army found that equipment deployed during recent operations often requires higher levels of maintenance during and following its time deployed. We model this by giving each item a base chance of requiring a trip to depot at the end of a possessing unit's time in theater, which we set at 50 percent for helicopters, based on discussion with SMEs. These items are then sent to depot for maintenance for about 100 ± ten days.
- **Unit overlap in theater.** Army units overlap in theater for a certain period of time in order to facilitate the transition between units, which creates additional demands for equipment. The unit

scheduling and deployment algorithm attempts to overlap units by two to four weeks, depending on the type of unit.

- **Dwell times.** Dwell times are derived as a result of demand signals and force structure and are a core output. Assumptions have to be made about deployment policies that affect the balance of dwell times between active and reserve forces. We have assumed that if a unit must be deployed early, active units in the Ready phase will preferentially be sent, followed by reserve units in the Ready phase, followed by active units in the Reset phase, and finally by reserve units in the Reset phase, with ties broken by the unit that is furthest along in any given phase.

Sensitivity Analysis

The inputs to REMM are flexible enough to accommodate a number of different demand and force structure scenarios. We have performed sensitivity analyses to demonstrate the effects of changes in combat brigade demand and active/reserve component mix on total required platform numbers, as well as the effect of changes in combat brigade demand on the dwell time of units. While a separate technical document will detail all algorithms in the model, along with a more complete sensitivity analysis, this section is meant to provide some additional context for the numbers provided in this report.

Demand Signal Versus Platform Numbers

Assuming a rotational-low equipping strategy (as detailed above), we have computed the number of each major end item required in our analysis as a percentage of the total currently in MTOE, TDA, and APS stocks for CABs and HBCTs. For CABs, we varied the demand signal ranging from a one-CAB steady state and two-CAB surge all the way to an 11-CAB steady state and 21-CAB surge, where the last scenario represents the simultaneous deployment of the entire Army for a one-year period. Likewise, for HBCTs we varied the demand signal ranging from a one-HBCT steady state and two-HBCT surge all the way to a 12-HBCT steady state and 24-HBCT surge scenario. The results of these tests are shown in Figures A.4 and A.5. While these scenarios are outside the numbers used in the analysis for planning, they provide an example of how the model responds to different inputs.

For both types of brigades, equipment requirements remain virtually constant up to about six CABs and seven HBCTs deployed steady state (roughly corresponding to the force structure of a 20-BCT steady-state demand scenario), because units in the Available phase require 100 percent of equipment regardless of deployment status, and the size of the Army's current force structure requires that in these scenarios some units are always in the Available phase, regardless of demands. However, in the more stressing scenarios above 6/12 CABs and 7/14 HBCTs, total equipment numbers required in an ARFORGEN policy begin to increase quickly as equipment needed to meet surge demands eventually overtakes steady-state equipment numbers for a constantly deployed force. These results are reflected in the graphs that follow.

Figure A.4
Major End Item Numbers Needed as a Function of Demand CABs

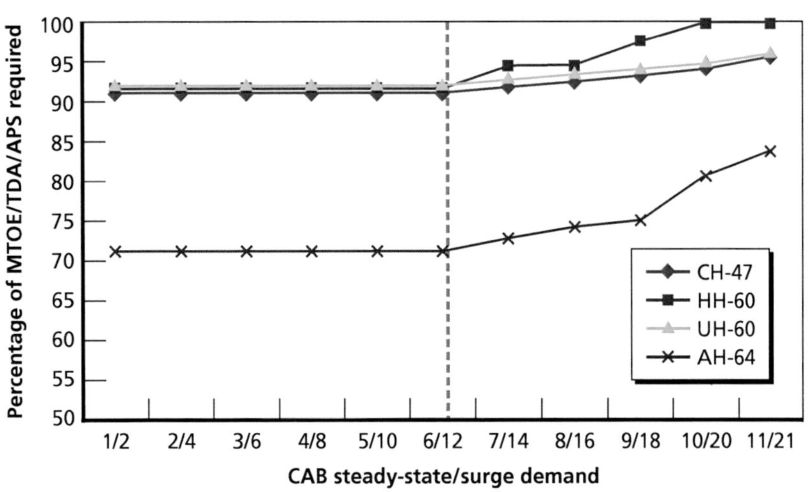

NOTE: Nominal 20/40 BCT scenario at blue line.

RAND *MG1092-A.4*

Figure A.5
Major End Item Numbers Needed as a Function of Demand HBCTs

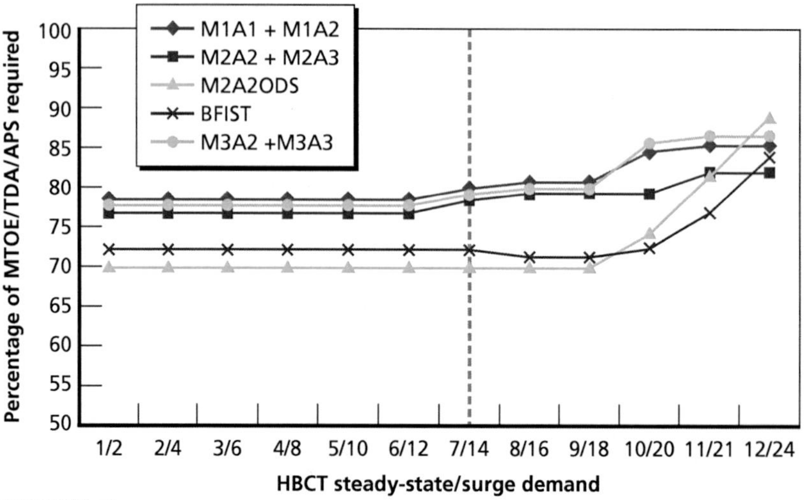

RAND *MG1092-A.5*

Active/Reserve Component Mix Versus Platform Numbers

We have similarly gauged the effect of changing the ratio of AC to RC combat brigades on major end item numbers. We again use a rotational-low equipping policy and a 20/40 BCT demand signal corresponding to six CABs and seven HBCTs steady state and 12 CABs and 14 HBCTs surge. As shown in Figures A.6 and A.7, as the number of AC brigades increases, the total needed stock of combat-only platforms (such as M1 tanks and AH-64 helicopters) increases, but the total stock of CDU items (such as Chinook helicopters) decreases.

Effect of Demand Scenarios on Combat Brigade Dwell Times

Because a core function of REMM's algorithms is to schedule and deploy units to meet theater demands, we can also gauge the stress on the force as a function of demand signals. As in our first test, we scale our demand signal from one CAB and one HBCT steady state and two CABs and two HBCTs surge all the way to 11 CABs and 12 HBCTs steady state and 21 CABs and 24 HBCTs for a one-year surge. We depict, in Figures A.8 and A.9, the average days deployed early per deployment, per combat brigade. Up to a "breaking point"

Figure A.6
Major End Item Numbers Needed as a Function of Number of AC CABs

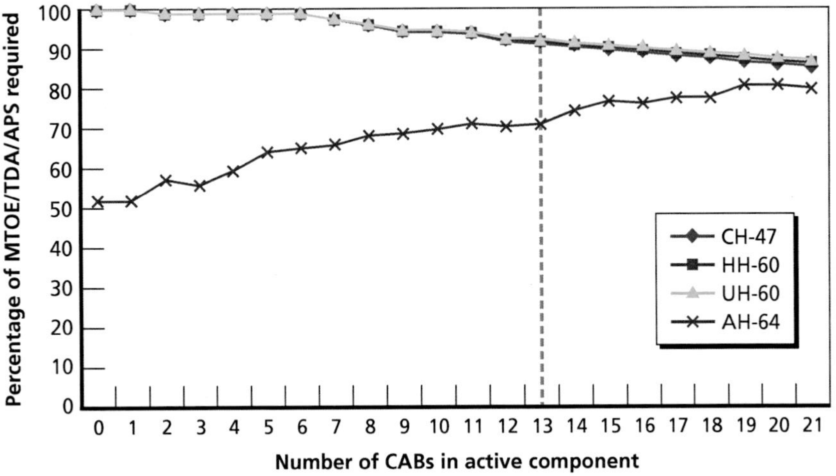

NOTE: Current force structure (13 CABs) is shown in blue.
RAND *MG1092-A.6*

Figure A.7
Major End Item Numbers Needed as a Function of Number of AC HBCTs

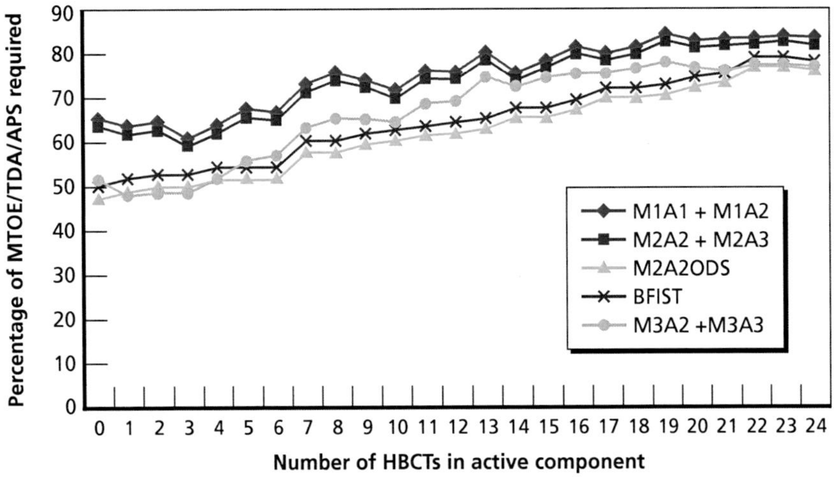

RAND *MG1092-A.7*

Figure A.8
Average Days Deployed Early per Deployment, per Combat Brigade, as a Function of the Demand for CABs

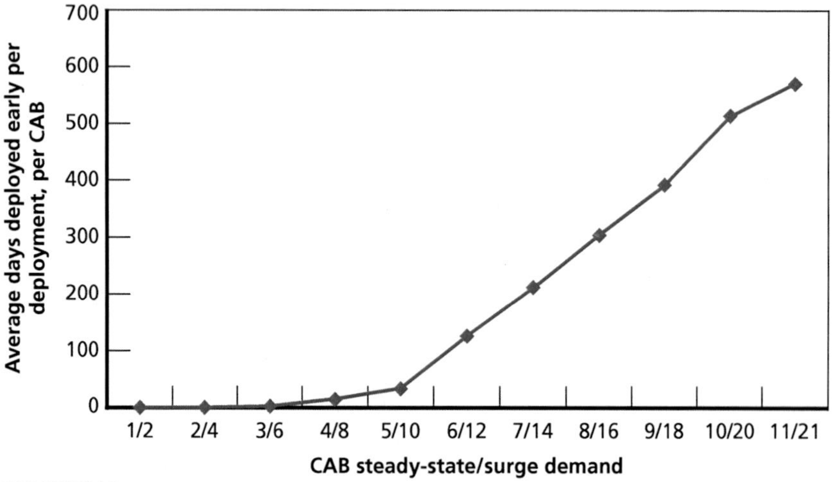

RAND *MG1092-A.8*

Figure A.9
Average Days Deployed Early per Deployment, per Combat Brigade, as a Function of the Demand for HBCTs

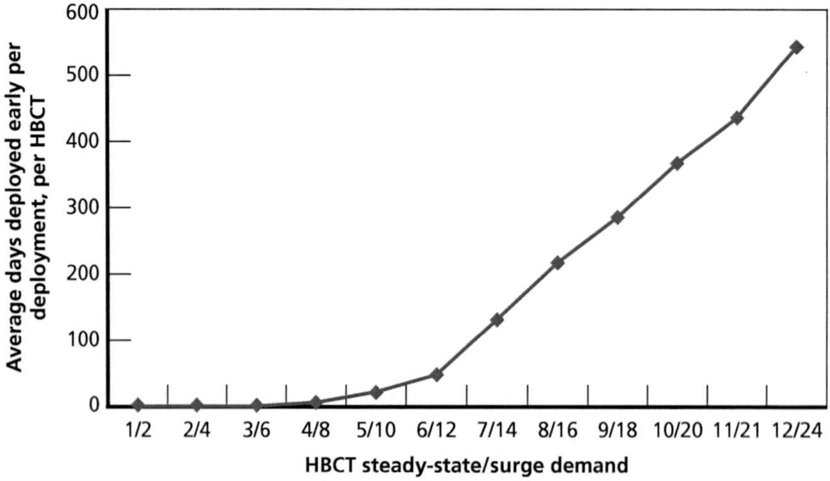

RAND *MG1092-A.9*

of five CABs and five HBCTs steady state, the force structure is sufficient to accommodate the demand signals such that the average days deployed early is about one month over a 15-year period. However, after this point, early deployment times quickly rise.

Bibliography

2010 Army Posture Statement, "Addendum F, Army Force Generation (ARFORGEN), the Army's Core Process," Army.mil, last modified March 2, 2010. As of September 24, 2010:
https://secureweb2.hqda.pentagon.mil/vdas_armypostureststatement/2010/addenda/Addendum_F-Army%20Force%20Generation%20(ARFORGEN).asp

Army Financial Management—Assistant Secretary of the Army for Financial Management and Comptroller, *Budget Materials—Fiscal Year 2011*, last modified November 11, 2010. As of February 17, 2011:
http://asafm.army.mil/offices/BU/BudgetMat.aspx?OfficeCode=1200

Bonds, Timothy M., Dave Baiocchi, and Laurie L. McDonald, *Army Deployments to OIF and OEF*, Santa Monica, Calif.: RAND Corporation, DB-587-A, 2010. As of February 11, 2011:
http://www.rand.org/pubs/documented_briefings/DB587.html

Carpenter, Raymond W., "Statement Before HASC on Army and Air National Guard Equipment," Washington, D.C., May 5, 2009.

Casey, George W., "The Army of the 21st Century," *Army Magazine*, October 2009, pp. 25–40.

Defense Acquisition Management Information Retrieval, *Selected Acquisition Report: AB3*, Washington, D.C.: U.S. Department of Defense, December 31, 2009.

Headquarters, Department of the Army, Logistics Information Warehouse website, undated(a). As of December 4, 2009:
https://liw.logsa.army.mil

Headquarters, Department of the Army, *FMSWeb (Force Management System Web Site)*, undated(b). As of December 5, 2009:
http://webtaads.belvoir.army.mil

Headquarters, Department of the Army, *The Army Equipping Strategy*, white paper, Washington, D.C., 2009.

Rotte, R., *Your Army's Commitment to Aviation*, Washington, D.C., November 2, 2008.

Schoomaker, P. J., "On the Army's Preparedness for Current and Future Missions," testimony before the House Armed Services Committee, Washington, D.C., January 23, 2007.

Solis, W. M., *Defense Logistics: Preliminary Observations on the Army's Implementation of Its Equipment Reset Strategy*, Washington, D.C.: United States Government Accountability Office, 2007.

U.S. Department of the Army, "Addendum E: Army Force Generation Model—ARFORGEN," *2006 Posture Statement*, site last revised February 8, 2006. As of February 2, 2010:
http://www.army.mil/aps/06/15_addendums.html

U.S. Department of Defense, *Conduct of the Persian Gulf War,* Final Report to Congress Pursuant to Title V of the Persian Gulf Supplemental Authorization and Personnel Benefits Act of 1991, Washington, D.C., April 1992.

U.S. Department of Defense, *Quadrennial Defense Review Report*, Washington, D.C., February 2010.